INFANTRY LIFE

INFANTRY LIFE

VIETNAM'S CENTRAL HIGHLANDS, 1966 – 1967

DENNIS M. WITT

Deeds Publishing | Athens

Published by Deeds Publishing in Athens, GA
www.deedspublishing.com

Printed in The United States of America

Cover and interior design by Deeds Publishing

ISBN 978-1-961505-06-3

Books are available in quantity for promotional or premium use. For information, email info@deedspublishing.com.

First Edition, 2023

10 9 8 7 6 5 4 3 2 1

This book is dedicated to my fellow infantrymen in Vietnam, whose daily lives as infantrymen filled their memories with experiences similar to those I've described in this book.

Contents

1. Background

My name is Dennis Witt. I served in Vietnam in the Army from October 14, 1966 until October 12, 1967. I was an infantryman (See Figure 01-1). These stories are about many of my memories of that year. My intent is to give the reader an understanding of what my life was like as an infantry soldier in my unit during that year. Several of these stories I've told many times when gathered with friends who were interested in hearing about some of my Vietnam experiences. Many of these stories I have not shared with anyone, ever, until now.

Some stories are funny—at least to me, some are sad, some are just memories of particular events that made an impression on me. Most are intended to describe what an infantryman's daily life was like in Vietnam. Almost all are directly from my own experiences. A few of the stories were told to me by the soldiers who experienced them. I included those secondhand stories because I thought they were interesting.

The stories are as truthful and honest as my memory can make them. I tried to never exaggerate. To make the stories more understandable to the reader, I did take the liberty of filling in small parts of the stories with 'what must have happened' or 'what usually happened' whenever my direct knowledge or memory was missing that information.

All the stories are designed to be told stand alone. That is why some things are described in stories that are also part of other stories. I felt those repeated things are important to know if one is to understand that particular story. During many of the stories, I've mixed in descriptions of some aspects of being an infantryman in Vietnam that might not pertain directly to that story, but I felt might be interesting to the reader. I tried to write each story like I would tell the story verbally while sitting around a campfire.

At the end of this book is a glossary of many of the acronyms and terms we used in Vietnam. Many of the terms are used repeatedly in these stories. Having the glossary available means I didn't have to constantly define and redefine the terms in each story. Please use it to translate the terms that are unfamiliar to you.

In Vietnam, I served in an infantry battalion whose nickname was 'The Red Warriors'. We were the 1st Battalion of the 12th Infantry Regiment, part of the 2nd Brigade of the 4th Infantry Division.

My year was divided into four different phases. I served as a rifleman in second platoon of Charlie Company for the first two and a half months. I was then one of Charlie Company's commanding officer's two RTOs for the next six months. Next, I served as the battalion commander's RTO for about a month. I finished my year working as an RTO in the battalion's Tactical Operations Center (TOC) for the last two and a half months.

The battalion had, while I was there, three infantry companies usually containing about 100 to 120 men each. Counting the three infantry companies and all the support troops, the battalion had about 500 to 600 active soldiers at any given time. Most of the time, our battalion was assigned a specific area of operation (AO), in which all our units worked together to accomplish the battalion's mission, whatever it might have been.

Except for a Christmas rest and one other five- or six-day rest, our battalion was out in the boonies for my whole year, spending almost every day looking for the enemy. The enemy we searched for was primarily the North Vietnamese Army (NVA). We usually operated close to the Cambodian border, along what was called the Ho Chi Minh Trail, generally west of Pleiku, Vietnam. We operated there because the NVA made their base of operations just on the other side of the border in Cambodia. While I was in Vietnam, we were forbidden from crossing the border in pursuit of the NVA, so Cambodia was a safe haven for them.

I believe the NVA's general policy for fighting the war in our AO was to fully engage US troops only when the NVA felt they had a significant advantage. If they did not have the advantage, they quickly disengaged and fled back into the depths of the jungle or into Cambodia. Because of that policy, the Red Warriors had very few large, face to face, engagements with the NVA. Anytime we encountered them, they almost always disengaged after a short fire fight, because in their opinion, we had the advantage—be it in men or firepower.

However, there were three times during the year I was in Vietnam when the NVA chose to stand and fight the Red Warriors. They thought they had the advantage.

The first battle happened just before I joined the company in the field, so I was not involved. The second battle, I was heavily involved in. Both of those battles were disasters for the NVA. They miscalculated and paid a heavy price. But, during the third battle, the NVA were able to inflict heavy casualties on one of our companies before they disengaged and fled back into Cambodia. I was not involved at all in the third battle, because I was out of country on R&R when it happened.

What I'm trying to emphasize here is, while my battalion was in a war zone, mostly in a free-fire zone in fact, heavy daily contact with the enemy just didn't happen. We were not in a battle for our lives every day. Sometimes, the battalion would go for a week or two without any contact at all. If there was contact, many times it only involved one of the three companies. And then maybe only one squad or platoon from that company.

Most of our days were spent looking for but not finding the enemy. I tell you this because I do not want you expecting the following stories are going to be a series of bloody combat stories. They are not—mostly because of what I've said above. While contact with the enemy was a real possibility every day, heavy contact just didn't happen that often.

To try to help you understand these stories and me better, I'd like to describe a little bit of what my life was like before I went into the service and what my mindset was like during that time in my life.

In the year just before I went into the Army, I was in college attending the University of Wisconsin—Kenosha Extension. I still wasn't sure what I wanted to major in, but I felt that getting an education was very important. I was a freshman and 21 years old.

I was that old because, shortly after graduating high school, in 1962, I decided that I wanted to join my dad and uncle in their family bakery business. I knew that, if I wanted when they retired, I might eventually own the business.

However, after 2 ½ years of working there, I very reluctantly decided that the small bakery business wasn't what I wanted to do for the rest of my life. There were many reasons that went into my decision, but for this book they are not important to talk about.

At that point in my life, one of the hardest things I ever had to do was tell my dad that I was going to quit the bakery business and go to college. I know it hurt him to learn I was not going to follow in his footsteps, but it was a decision he eventually supported.

So I went to college. I started in January of 1965. During 1965, the war in Vietnam was really ramping up. In fact, during that year, a good friend of mine was in Vietnam with the 101st Airborne Division.

Politically, I was a supporter of why we were in Vietnam. I believed, and still do, that the spread of communism had to be stopped. I believed in what they called, 'The Domino Theory'. That was the idea that communism would spread from country to country, like dominos falling, if it wasn't stopped somewhere. I thought making our stand in Vietnam would help prevent the rest of southeast Asia, and the countries around it, from falling under communistic rule. I was what was known as a 'hawk' when it came to the war.

As my year in college progressed, public opposition to the war began to increase. Because men were deferred from being drafted while they were in college, guys were constantly being accused of attending college simply to avoid the draft. I knew that was not why I was in college, but it bothered me that others might think that of me.

Finally, at the end of my freshman year—in mid-January of 1966—I made another major decision in my life. Since I still had no clear idea of what I wanted to major in, and because I was a patriot, I thought I'd quit college for a while and do what I felt was my duty to my country—go into the Army. I also thought that, perhaps after a couple of years in the Army I'd have a better idea of what I wanted to do for the rest of my life.

Therefore, I quit college and did an even harder thing—I told my parents I was going into the Army. It took until I was a parent myself for me to fully realize what they must have felt about that, knowing there was a war going on and I might eventually be a part of it.

In 1966, if you walked into an Army recruiting station and signed up for the Army, it was a three-year commitment. However, if you were drafted, it was only a two-year commitment. At 21, I didn't feel I wanted to spend three years in the Army. I thought I could accomplish what I wanted to accomplish by just serving two years. So, I decided being drafted was the best way for me to go.

I didn't want to wait around for the draft to work its very random selection process. Therefore, in mid-January 1966, I went to the Selective Service office in Kenosha and signed up for the draft. That meant, after processing was done, I'd soon be getting my draft notice. 'Soon' to the government, was the middle of March. That is when I finally got my notice that I was going to be inducted into the Army on April 14, 1966.

After wrapping up my affairs at home, I entered the Army at the Milwaukee induction center. From there I, and probably 30 other guys, got on a train and traveled to Fort Knox, KY for basic training. We traveled in a Pullman car. I was pleased with that because I got to experience sleeping on a train, in a bunk-like bed, like so many people in the past had done when trains were a much more commonly used form of cross-country travel. Even in 1965, that was becoming a rare experience for people. Very few ever do that now. I thought it was cool.

During basic training, they always give you various sales pitches for jobs you can volunteer for in the Army. Their sales pitch usually included an invitation to 'sign up for more information'. I had already decided I wanted to be a paratrooper — I was in the Army anyway and I wanted to get as much out of it as possible — so I signed up for 'more information' about being a paratrooper. I convinced two of my best friends in basic to sign up for 'more information' too. What could it hurt?

After finishing two months of basic training, I got a one week leave, and then went to Fort Gordon, GA for Airborne Infantry — Advanced Infantry Training (AIT). When I got there, I learned my two buddies were also there. Everyone in this school was expected to follow it up with a trip to Jump School. So much for just 'more information'.

After completing AIT, we all bused to Fort Benning, GA for Jump School. After graduating Jump School, I was excited to now be a hard ass paratrooper. I went back home for a 10 day leave and then started my journey to the next year in my life — in Vietnam.

After my tour in Vietnam was over in October of 1967, I was stationed at Fort Carson, CO for my last five and a half months of service. After a year in tropical Vietnam, I got to spend the winter in Colorado. Nice.

At Fort Carson, I was in the Fifth Infantry Division, which was a mechanized division. That meant that the infantry soldiers mostly rode in armored personnel carriers (APCs) instead of walking. My whole unit was comprised of enlisted men who were just like me, Vietnam veterans just waiting to complete their time in the Army. Our main job, as I saw it, was to train new second lieutenants with their first command experience before they went to Vietnam.

While there, I was trained as an APC driver, which was fun. We played 'war games', slept out in the cold and snow of a Colorado winter a few times, and generally had a pretty easy Army life for five and a half months.

Just before getting out of the Army, I was promoted to buck sergeant. Not bad, I thought, making E-5 in less than two years.

I know I was a good soldier and am very proud of that. I was awarded paratrooper's wings when I graduated Jump School after successfully jumping out of a perfectly good airplane, five times. In Vietnam, I was awarded the Combat Infantryman's Badge (CIB). The CIB is only awarded to infantrymen who have directly engaged the enemy and have performed their duties satisfactorily.

I was also awarded two Bronze Stars, one for Valor and one for Service. The Bronze Star for Valor was awarded to me for my actions during the Battle of 501N in February of 1967 (See Figure 01-2). The Bronze Star for Service was awarded to me for my performance as the RTO for the Company Commander of Charlie Company, my performance as RTO for the 1/12th Battalion Commander, and for the job I did in the Tactical Operations Center for the 1/12th. I am proud to have been so honored by my fellow soldiers and my country.

After I left the Army and looked back at my year in Vietnam, I realized I was very, very lucky. Much luckier than the average infantryman.

I have always felt that there were many things that happened there that might have really bothered me for the rest of my life. But, because of luck or, maybe God, I was not physically present for many of those events. Almost every time something really bad happened in or to my unit, I was not there. Not be-

8

cause I didn't want to be or tried not to be, but because of circumstances. I have memories of those events, but not firsthand
memories, just the memories of what I was told by others.

I was never wounded. I never saw any of my friends badly
wounded. I never witnessed anyone die. I never carried bodies.
I was never involved in any civilian casualties of any kind. I may
have killed several enemy soldiers, or I may not have—I really
don't know.

All I do know for sure about my service in Vietnam, is that
I was very fortunate and did not come back with any serious
PTSD. For that I have always been very grateful. I have many
friends who were not so lucky.

Again, the following stories are meant to give the reader an
idea of what life was like for me there, what it was like for all of
us infantry soldiers in the Red Warriors. These stories are both
about life in general for soldiers in the Red Warriors, and about
specific things that happened to me and to my unit that made
our lives what they were.

Many veterans of my unit share some of these memories
with me, but at the same time, they also have a completely different set of memories of that time as well. Or they may remember some of the shared experiences in a different way. I've
learned by talking to the guys at our reunions that our memories of that time are varied. It all depends on what experiences
made an impression on individuals as they went through their
time in Vietnam.

Some veterans of other units in Vietnam may have had
similar experiences to mine. More likely, though, is most other
veterans probably had a completely different set of experiences
in Vietnam. Their daily lives were probably completely different from mine. Soldiers' experiences always depended on their

9

MOS (their military job), the unit they were in, the part of the country they served in, and when they were there. I've met other infantry vets who served in other units, in other parts of Vietnam, whose experiences and lives in Vietnam were significantly different than mine.

These are my stories and my experiences as I remember them.

I've always described the year I spent in Vietnam as the most significant year of my life. It helped make me the man I am for the rest of my life. It was a year I would never want to do again, but, at the same time, it was a year I would never trade out of my life for anything.

I hope these stories help you understand what that year was like for me and for the rest of us in the 1/12th, Red Warriors, in Vietnam.

Figure 01-1: This picture is of me sitting on a perimeter bunker at a firebase. The bunker looked out over the firebase LZ. A Chinook had just made a delivery and was taking off. A buddy of mine took a perfectly timed picture of me and the Chinook with my camera. My mom always told me this was her favorite picture of me in Vietnam.

Me and the Chinook

Figure 01-2: These were the Charlie Company soldiers who were awarded medals, primarily for their actions at the battle of 501N. Our 1st Sgt is on the right and I am in the center. Daniel, the wounded soldier who shared the bunker with me that day, is between me and Top.

Awards Ceremony at Firebase

2. Flight to Vietnam

My best friend, Jim, volunteered to drive me from my parent's home in Kenosha, WI to O'Hare field in Chicago where I would start my journey to Vietnam. It was a beautiful October day and my family (See Figure 02-1) and I were all saying goodbye in front of my parent's home.

As I hugged and kissed each of them goodbye, I actually thought it was for the last time ever. I really believed I would probably lose my life in Vietnam. But I was OK with that. I felt going there was my patriotic duty and if I didn't come home again, well, that was God's plan. However, I didn't mention those feelings to my family.

Jim had been out of the Army for about a year. During his time in the Army, he had spent a year in the infantry in Vietnam, so he knew about some of what I was going to experience in the next 12 months. On the drive to O'Hare, he tried to give me pointers on what to expect. Being as naïve as I was, most of his advice didn't really stick very well in my head.

When he dropped me off at O'Hare, we shook hands, then hugged, and I was on my way.

I flew from O'Hare to an airport near Fort Dix, NJ and reported in at Fort Dix. After processing, a large group of other soldiers and I destined for Vietnam were bused to a nearby Air Force base. There we all got aboard a large C-141 cargo transport jet.

The C-141 was set up to transport troops. The rear portion of the fuselage was the cargo area. At the end of the fuselage was a giant door that pivoted down and became a ramp for unloading the cargo. All our duffle bags were stored in that cargo area.

The inside of the rest of the fuselage was set up with seating. It was not the kind of seating you'd see on a commercial passenger jet, but the kind of seating used for troops. There were bench-like seats against the outer walls along the entire length of the fuselage. The seats were made of nylon mesh. Down the whole middle of the aircraft was an additional row of back-to-back bench seats that were made of the same nylon mesh. The arrangement of these seats was almost identical to the seating in the C-130's I jumped out of in jump school. This was the set-up used by paratroopers. Unfortunately, the nylon bench seats were not very comfortable to sit in, especially for the long flight ahead of us.

We probably had well over 100 troops on that plane going to Vietnam. But, thankfully, the plane was not packed to capacity, so we were not crammed into the seats butt to butt. That meant there were some spots where you could stretch out a bit. You could also walk up and down the aisles to stretch your legs occasionally.

Flying in that C-141 had one real drawback. There were only two small round windows we could look out of, one on either side of the fuselage just in front of the cargo area in the back. No sightseeing was available as you flew along, unless you made the trip back to those small windows. When you did, you usually had to share viewing time with a couple of other guys.

To start my journey, I had flown east from Wisconsin to get to New Jersey. When we took off from New Jersey in the C-141, we flew back west. The captain of the plane had an intercom system and made announcements concerning our flight as we went along. He periodically told us where we were and what time we were scheduled to reach our next destination, which was nice to know. Just after we took off, he told us our first destination was an Air Force base in Alaska.

At that point, I had not done much flying in my life. All my commercial flights thus far were in the Midwest to and from the Army posts where I trained. I had never flown out west before.

A few hours after we took off, I made my way back to the two windows and took a peek. Amazing! We were flying over the start of the Canadian Rockies. Everything was snow covered. Spectacular view. It was the first time I had ever seen mountains. I watched the landscape go by under us for several minutes until I noticed a couple of other guys were waiting for a look. Reluctantly, I went back to an open seat.

Several hours later, we landed in Alaska. The scenery at that Air Force base was awesome. Tall, snow-covered mountains almost surrounded it.

We taxied on the tarmac and stopped near some buildings. They had us deplane through two doors on the left side of the aircraft, both with stairs or gangplanks. Out on the tarmac we had to wait a few minutes for buses to come to get us. It wasn't extremely cold that day, but for us, it felt very cold. We were all wearing our Khaki short sleeved uniforms, which would be great in Vietnam, but were very impractical in Alaska in October.

The buses arrived and we were taken to a large, very nice, mess hall where they fed us. It was then we Army guys discovered that Air Force guys had it really nice. We had maybe an hour or so to eat and then were bused back to the plane. It was just before sundown when we took off, heading west again.

The next few hours were interesting to me because sundown and twilight lasted so long. We were flying west, chasing the sun. It took a long time before we fell far enough behind the sun that it finally became dark. I spent as much time as I could looking out one of those two small windows until it was too dark over the ocean to see anything.

Our next destination was Tokyo, Japan. We landed at Tokyo airport in the dark and taxied to an area way off to the side of the runways. There was some delay in our leaving, so they let us off the plane to stretch our legs a bit. I think we were on the Tokyo tarmac for about an hour before they told us to get on-board again. We took off and just a short time later we landed at Yokota Air Base, still in Japan. Once there, we again deplaned and were bused to another mess hall for another meal. There was no meal service on a C-141.

After the meal we sat around for one to two hours and then got back on the plane. We took off and the next stop was announced as Saigon, Vietnam. It was still dark when we took off from Yokota, but soon the sky started to brighten up a bit.

Early in the flight, the captain of the plane had invited us to individually come up to see the flight deck if we wanted. I finally took my opportunity just as we approached the coast of Vietnam. The view out the front windows of the plane was amazing. There in the distance was the green, green coastline of

Vietnam with a deep blue ocean in the foreground. It was hard to imagine something so beautiful was the home of a savage war.

We landed in Saigon and waited anxiously for the doors to open. The first door to open was the rear cargo door. As it opened, very hot, humid air rushed in. Now our khakis were appropriate. The whole flight from New Jersey to Saigon took 33 hours.

When the side doors opened and we began to deplane, I was immediately struck by what we saw on the tarmac. We were parked in an area where many fighter planes were haphazardly parked everywhere. Surrounding each fighter on three sides were high walls of sandbags. There were many military personnel walking around, all armed with weapons. Sandbag bunkers were scattered around the area as well. It was obvious we had just arrived in a war zone.

Buses were sent to our plane. After finding and retrieving our duffle bags from the cargo area of the plane, we boarded the buses which would take us to the replacement depot on the outskirts of Saigon.

When we got onboard the buses, the first thing we noticed was that the windows on our bus had wire screening covering the outside. One soldier aptly pointed out that the wire was probably there to prevent grenades from being thrown through the windows. It was very unsettling riding in that bus because we were all unarmed. Besides the bus driver, I think there were only one or two armed soldiers onboard for security.

Driving through the crowded streets of Saigon was another first for me. I had never been outside the US in my life. This city was unlike any city I had ever seen. Narrow, crowded streets, poor looking inhabitants, and the concern that any one of them might be a VC plotting how he could kill us. Again, very un-nerving.

After reaching the replacement depot and a few days of pro-cessing, I officially became, a soldier in Vietnam.

Figure 02-1: My family. Taken shortly before I left for Viet-nam.

My Family

3. Not in US Anymore

It was mid-October, and my first day in Vietnam. I had just landed at a military airport near Saigon aboard a C-141 transport plane after a 33-hour trip from New Jersey. After arrival I, and the rest of the plane load of GI's, were bused to a huge replacement depot just outside of Saigon. The area around the repo depot, as we called it, had roads, some businesses, and houses. The civilian area was much more spread out than in crowded Saigon, which we had passed through on our way from the airport to the repo depot.

After arrival and initial processing, we were all assigned to our barracks. The barracks were large wooden buildings, filled with bunk beds. We new arrivals were mixed in with guys who had been in the repo depot a few days and were still awaiting their assignment to a permanent unit in Vietnam.

When I found an unused bunk, I put my duffle bag on it and sat down. The guy in the next bunk and I started talking. He said he had been in the repo depot for almost a week. He wasn't sure why it was taking so long to get him assigned to a unit, but he wasn't too anxious to leave. He found life was not too bad there. I asked him what I should expect.

He told me that, typical of the Army, during the day while you were waiting to be assigned, soldiers were given duties like KP, police call, etc. Most of the duties were not too fun. He said if I wanted to avoid those crappy details I should volunteer for guard duty on the perimeter of the repo depot. He was told this area of Saigon was very secure and the chances of an attack on the repo depot were remote.

The next morning, I volunteered for guard duty and was ordered to report for an orientation. There they explained the duties I would have while standing guard at one of the large, heavily sand bagged bunkers on the perimeter of the repo depot. I was told I was never to fire my weapon unless actually fired upon. I guess they didn't want any trigger happy GI's shooting up the civilian areas just outside the perimeter for no good reason.

Just before my shift started, I was given a steel pot, an M-14, and a couple of magazines of ammo. I was then driven to one of the perimeter bunkers. The Repo depot had a wire fence surrounding it. The fence was maybe 30 meters in front of the bunker and the area between the fence and the bunker was clear of any vegetation. A great field of fire.

Just 20 meters beyond the fence was a busy roadway with lots of traffic. Between the road and the fence was a sidewalk or path that occasionally had people walking on it.

I was by myself in the bunker. The substantial bunker was built out of sandbags and logs. It was totally above ground and had overhead cover. It was like a big box built of sandbags. There was an entryway in the rear. The front wall was a little over waist high and had a three-foot-high horizontal opening across the whole front. Above the opening was a big log from wall to wall. More logs went sideways across the top of the bunker, supporting the sandbagged roof. You could stand up in the bunker and lean your elbows on the front wall as you watched to the front.

The first hour or so was uneventful. Traffic on the road and occasional pedestrians walking beside the road was all that went on. It was interesting watching the pedestrians. Many were Vietnamese women. Most of them were petite and slender. They all wore the traditional long colorful dresses of Vietnam. Some had pretty umbrellas.

Then something happened that helped me realize I was not in the US anymore. As I watched the area in front of the bunker, I saw two women walking beside the road, approaching from the left. At the same time, I saw a man approaching from the right, also walking beside the road. Then something unbelievable, at least to me, occurred.

As the people approached each other, the man stopped, turned to his left (toward me), took one step, and began to unzip. The women were still walking toward him, maybe 20—30 feet away. The man began to pee! The women just continued walking and went past him without batting an eye as he continued doing his thing. When he was done, he zipped up and continued on his way. The women seemed unfazed.

But I was shocked. Never had I expected something like that to happen. I realized then that customs here in Southeast Asia were significantly different than they were back home. I knew then I was in for a year of new experiences that were far out of the ordinary for me. I was oh so right about that.

I volunteered for guard duty over the next few days. Then, I was called in and told what my new unit would be. Since I had graduated jump school just a few weeks before, I was fully expecting to be assigned to an airborne unit in Vietnam. The 101st Airborne Division and the 173rd Airborne Brigade were both in country at the time. The officer told me I was being assigned to the 4th Infantry Division (4ID).

I quickly pointed out that the 4th was not an airborne unit, and I was a paratrooper. He shook his head and said that the airborne units didn't need replacements at this time, but the 4th did. So, I was going to the 4th Infantry Division. He then dismissed me.

I was very upset about that, but, in the end, it turned out to be a good thing for me. The 1/12th Infantry Regiment in the 4ID was a very good unit. It had a lot of fine men, many of whom became my close friends. I am very proud to have served in the Red Warriors throughout my time in Vietnam.

4. First Night Smoker

Late October, 1966. It was my first day in the 4th Infantry Division base camp located near Pleiku, Vietnam. I had spent most of the day traveling there from the repo depot near Saigon. My modes of transport were bus, cargo plane, and deuce-and-a-half truck.

When I got to the 4ID base camp, I was processed in and assigned to Charlie Company, 1st Battalion, 12th Infantry Regiment. The 1/12th was known as 'The Red Warriors'. I was taken to the Charlie Company office where I checked in and was assigned to one of the large tents that served as barracks. At that time, the 4ID base camp was almost entirely composed of large tents. There were very few wooden buildings yet. I found an open cot and put my duffle bag on it.

Soon I met a couple of the guys. Since it was late afternoon, we all went to the mess tent for some food. After we ate, the guys took me to a tent that served as a sort of an enlisted man's club. It was just a tent with some chairs and a few tables in it. There may have been some vending machines, but I don't remember.

In the middle of the tent was a circle of chairs. There were maybe 15 to 20 chairs in the circle. Several of the chairs were occupied already. The guys I was with, and I sat down in three that were empty. The circle was just being used for a BS session.

Soon most of the chairs were filled up and several conversations were going on at once. Since I was the new guy, I pretty much kept my mouth shut and just listened. I don't remember much of what was talked about.

A bit later in the evening, a guy directly across the circle of chairs from me, lit up a cigarette. I noticed that the cigarette kind of looked like the ones my grandpa smoked. The ones he rolled himself.

The guy that lit that cigarette took maybe two deep drags and then passed it to the guy on his left. Hmmm, they share their cigarettes here, I thought. Then the second guy took a drag and passed it to the third guy. Hmmm, cigarettes must be really hard to get here, they are all sharing one cigarette.

When the cigarette was passed to the fourth guy, my thoughts suddenly began to change. What if it's not a cigarette? What if that is marijuana? Drugs! Oh my God. Narcotics, right here.

I was a very naïve kid from Wisconsin. I had never been exposed to pot in my life. The only thing I knew about it was what I learned in a high school health class. There, they told us marijuana was addicting and a terrible gateway drug to much harder drugs.

This was bad, I thought. I didn't want to get addicted. What should I do if that marijuana cigarette makes it to me? What will these guys do if I don't smoke it?

Just then, the fourth guy offered it to the fifth guy. He just shook his head no and the fourth guy simply passed the cigarette to the sixth guy. No one said anything to the fifth guy. Thank God, you can refuse without apparent consequences.

When the cigarette got to the guy next to me, he took a drag. Then he offered it to me. I casually shook my head no and he leaned over and passed it to the guy on my left. Wow. I made it. No addiction for me.

When I joined the company in the jungle a week or so later, I learned that no one smoked pot out in the field. No one wanted themselves or their fellow soldiers to be high if we were attacked. That attitude held true during my whole time with Charlie Company.

Later in my year, at the firebase, it was a different story. Pot started being used there a couple of months before I came home. A few guys regularly 'partied' each evening in their bunker. They always asked me if I wanted to join them, but I passed. They didn't have any problem with that.

During my year, I didn't see or hear of anyone using anything harder than pot. However, that did not last in Vietnam. Unfortunately, harder drugs were more commonly used by soldiers in Vietnam in the later years of the war. Probably not so much by the infantry soldiers out in the field, but by the guys in the more rear areas.

5. One of the Guys

I arrived in Vietnam in the middle of October, 1966. After spending several days in the replacement depot outside of Saigon, I was assigned to the 4th Division. Specifically, to Charlie Company, 1st Battalion, 12th Infantry Regiment.

After another eight to ten days of orientation and training at the 4th Division's base camp, I was finally sent out to join the company in the field. It was late afternoon when we flew out. I was the only passenger on the helicopter as it traveled from base camp to the jungles of the Central Highlands. Charlie Company had been operating there for several weeks.

Before getting on the helicopter, I was given a large, black, plastic suitcase and told to give it to the first sergeant when I got to the company. I was also told to take care of it because it was worth about $6,000! At that time, nice new cars sold for $3,000.

As I rode in a helicopter for the first time in my life, I was amazed at the thick green jungle and tall hills everywhere below us (See Figure 05-1). There were no breaks anywhere that I could see. I kept wondering how they were going to get me on the ground.

Soon we started to circle something and also started descending. From where I was sitting, I could not see what we were circling. All I could tell for sure was we were quickly approaching the top of the dense, unending trees.

Suddenly, we were descending into an oblong clearing whose floor was littered with fallen trees and tree stumps. My first LZ. The helicopter stopped descending and hovered about six feet off the ground and the door gunner indicated I should jump out, NOW. So out I went with my M-16 in one hand and a $6,000 black plastic suitcase in the other. As soon as I hit the ground, I quickly made my way through the fallen trees and branches to the edge of the clearing and looked back to see what was going to happen next.

As I watched, the helicopter slowly did a 180 degree turn, then began to back up until its tail rotor was almost hitting the trees on the edge of the LZ. I didn't know helicopters could back up. Then it tipped forward and started a headlong charge toward the trees at the far end of the LZ. Just before crashing into the trees, it nosed up and climbed almost straight up like a jet. I didn't know helicopters could do that either.

As the sound of the rotors disappeared into the distance, all I could hear were the chuckles of the soldiers around the perimeter laughing at the apparent close call of a crashed chopper in their LZ.

I looked around and every soldier I could see looked very 'well used'. They had been out in the jungle on their search and destroy mission for several weeks without a break and looked it.

I asked the nearest guys where I could find the First Sergeant. They pointed the way. As I made my way to the First Sergeant, I got a few funny looks when they saw I was carrying a suitcase. I wonder what they all thought about one of their first 'new guys' and his suitcase.

I found Top, told him who I was, and gave him the suitcase. He was expecting it and muttered a few words about how big it was. He opened it right away, probably to see if the contents were OK. The suitcase contained an early version of a Starlight Scope — a night vision device — which looked like a large telescope. I guess they wanted to see if a Starlight Scope would be useful for detecting enemy soldiers sneaking up on the company in the middle of the night.

After a quick inspection of the Starlight Scope, the first sergeant sent me over to find my platoon leader. I found him and he then introduced me to my new team.

After the introductions, my team leader began briefing me about life in the boonies, how things were done, and what was expected of me.

As darkness approached, he explained that one guy from our team had to be on watch throughout the night. He said since there were four of us sharing a foxhole, we would divide the night into four watches of two hours each. I was the new guy so he said I could have the first watch.

At that time, we were not building hooches or using air mattress. For sleeping, we just laid our poncho liners on a cleared-out spot on the ground (well soaked with insect repellant of course), lay down on half of it, and pulled the other half over us for a little warmth. You usually needed that light blanket because the evenings and nights were often kind of cool in Vietnam in November.

It was a very rocky area and there were only three good spots to lay down behind the foxhole. My squad leader told me when my watch was done to wake my relief and then just lay down in his spot to sleep. Sounded like a good plan to me.

So, I spent my first two-hour watch in Vietnam, sitting next to the foxhole staring into the absolute darkness under the triple canopy jungle, contemplating all I had seen that day. Wow, for sure, this wasn't Wisconsin anymore.

After my two hours were over, I woke the next guy for his watch. He tossed back his poncho liner, got up, and went over to the foxhole. I laid down on his poncho liner and pulled it over me.

Oh my God, the stench of body odor immediately overwhelmed me. It almost made my eyes water. These guys had not bathed or showered for weeks and their clothes and, now I learned, even their poncho liners, reflected it. The only way I could sleep that night was to just cover my lower body with the liner and tough it out. Which I did.

After 10 days with the company, humping the hills to the point of exhaustion every day, I was beginning to feel a lot more comfortable with the life of being an infantryman in the Central Highlands of Vietnam. The evening of the 10th day, we ended up in a situation just like my first night with the company—just three sleeping spots and I had first watch.

When my watch ended, I woke up the next guy and lay down on his poncho liner. I pulled it up over me and about 10 seconds later I began to laugh to myself. I laughed because I realized I didn't smell anything this time. That meant that I smelled as bad as everyone else!

It was at that moment that I began to think of myself as 'one of the guys' of Charlie Company (See Figure 05-2).

Figure 05-1: This is a typical view of the Central Highlands west of Pleiku, Vietnam. This is the area that the Red Warriors operated in during much of the year I was in Vietnam.

Central Highlands

Figure 05-2: Several weeks after joining Charlie Company, this was taken at a firebase. I'm not sure, but these are probably the same fatigues I wore throughout that time.

One of the Guys

6. B-52 Strike

It was the morning of my first full day with Charlie Company. I had joined them the previous afternoon. We were packed up before dawn and were ready to move out at first light when word came around that we had a mission today that was unusual for the company. We were to go to the area of a B-52 bomb strike to check out whether it had effectively hit the enemy and, if so, determine what damage was done. They told us the strike target was designated for an area two kilometers from us and was scheduled to occur just after dawn.

When we were told that, several of the veteran guys around me were concerned that two kilometers was not very far away. They hoped that we were actually where we thought we were because a small error in either the B-52's targeting or our supposed location could be very bad for us. As a newbie, I could only take their word for what was going on and hope for the best.

We waited, ready to move out as soon as the strike was made. Suddenly, the bombs started exploding in the distance. There was a whole series of them. Probably 20 or more. When I say, 'in the distance', I'm exaggerating. They sounded like they were just outside our perimeter. As each bomb exploded, very strong concussion waves thumped at our chests and the trees around us actually shook.

It was all over in less than a minute. When it was, there was lots of nervous laughter at how close the bombs really were. Then, we started to move out.

As I said, this was my first morning with the company so walking through the thick jungle for the first time was, in itself, a new experience. I was amazed.

Traveling, as was normally the case in the jungle, was very slow. It was probably early afternoon before we covered the two kilometers to the strike area. My first inkling that we were close was when I looked ahead and saw we were coming to an open area in the jungle where light was coming through.

As we entered the cleared area, which was maybe 75 meters across or more, my very first thought was that this must be where the whole bomb load struck. All the trees in the clearing were almost devoid of leaves. Small trees and branches littered the ground everywhere, which made walking even more difficult.

As I made my way through the tangle of branches, I suddenly had a new realization. Ahead of me was a single crater. It was maybe 30 to 40 meters across and four or five meters deep (see Figure 06-1). I now understood. The clearing I was in was made by a single bomb! Ahead of us, through the thick jungle, I could see the edge of another clearing just like this one. There was one clearing for each bomb dropped. Awesome.

But there was one exception. As we searched through the area following the path of the bombs, I saw the soldier who was about 15 meters ahead of me stop and look down. Then he quickly moved on. As I approached the spot where he had stopped, I saw what he looked at. It was a hole in the ground maybe one to two meters wide that went down into the earth maybe three or four meters. The hole had a slightly curved path. At the bottom was just dirt. It didn't take a lot of thought to realize that under that dirt was a large bomb that has failed to explode. After a short look, I too thought it smart to move on as quickly as I could.

We found no sign of the enemy in the area of the bomb strike that day. Apparently, it was bad intel or a bad guess. No enemies were present. However, it was obvious if the enemy had been there or even relatively close by, they would have suffered many casualties.

We never again had another experience with B-52's like that. But this event added significantly to what turned out to be quite a first day for me in the jungles of Vietnam.

Figure 06-1: In the foreground is an example of a B-52 bomb crater. We built a firebase around this one. The crater is probably 30 to 40 meters across and four to five meters deep. The picture is of an air attack on the NVA that had mortared the firebase a short time before.

B-52 Crater Example

7. Vietnam — Search & Destroy formation

When Charlie Company moved through the jungle on a search & destroy mission, we tried to move in a formation something like what is shown below. The point platoon (1) would occupy the lead portion of the middle three columns. The left flank platoon (2) would be in the left three columns. The right flank platoon (3) would be in the right three columns. The company's command group (C) would be in the middle of the center column. The weapons platoon (W) would be in the rear of the center three columns.

8. Exhaustion

Before leaving base camp, I was given my basic complement of equipment. I had my ruck sack, which was a nylon pack attached to an aluminum backpack frame. I had my M-16, three or four magazines of ammo, and a couple of bandoleers of extra ammo still in their cardboard boxes. At that time there were not enough magazines available, so each man could not have all his ammo already loaded into magazines.

I was given a poncho and poncho liner (a light nylon blanket).

I was also given webbing equipment, which was a belt with a suspenders like attachment on which we carried much of our equipment. On the webbing I had an individual first aid bandage, some ammo pouches, several canteens, and two or three grenades.

All of the above, with a few personal items, was the extent of the equipment I carried that first day with the company. That is, until just before we started our march that morning.

After we finished packing up, my squad leader came over with a metal box of M-60 machine gun ammo and said I should carry that as my part of the squad equipment. (Later on, our machine gun ammo was still carried by members of the squad, but it was in belts of ammo draped over our shoulders, not in an ammo box). I think the metal box contained 200 rounds of ammo. It probably weighed 15 pounds or so. After getting my bonus equipment, we started out on our march.

That morning, I was only five or six weeks out of jump school where I was in the best physical condition of my life. Since graduating jump school, I had gone home for a 10 day leave and was probably in Vietnam for a little over three weeks. During that time, I had not done a lot of PT, so maybe I had fallen somewhat out of shape from my peak in jump school. But I was still in pretty good shape.

However, I was not at all ready for the physical stress I was introduced to that day on my first march through the Central Highlands. It was hot. It was humid. Walking up and down hills through the thick jungle, tripping over vines and logs, crawling under bamboo, and fording streams were all activities my body and mind were not used to. There was also the stress on me, as a new guy, wanting to make a good first impression. Added to that, was the worry of thinking that the jungle might also be filled with enemy soldiers who were very anxious to kill me.

Carrying that ammo can was also so damn awkward. With an M-16 in one hand and the ammo can in the other, balance was very difficult to maintain. I'm sure I went face down at least a couple of times that day. Probably much to the amusement of any of the guys who witnessed it.

As the day went along, every guy I saw looked like he was just out for a walk in the park compared to me. I was getting embarrassed. Once I even caught myself actually hoping someone would shoot at me so I could lay down on the ground and rest. The smarter half of my brain quickly pointed out how stupid that thought was, but the thought still lingered a bit.

I got a bit of a break when we slowed our pace searching through a B-52 bomb strike area to see if it had damaged the enemy. It had not.

Finally, about midafternoon we were done marching for the day and stopped to build our night perimeter. I was extremely relieved to stop because I was approaching total exhaustion. I can't remember for sure, but I probably had to dig the foxhole that night because I was the dumb new guy and that was about all I was qualified to do.

I was occasionally near total exhaustion a few more times during my year in Vietnam, but none of those times were so memorable as my first day in the jungles of the Central Highlands.

9. First Listening Post

It was my third day with the company in the jungles of the Central Highlands.

I was still trying to adjust to my new life as an infantryman in Vietnam, both physically and mentally. Physically because the terrain, the load I carried, and the heat and humidity of the jungle, were all stressing my endurance. I thought I was in good shape when I arrived in Vietnam, but the strain of this new kind of physical activity was very hard to get used to. Mentally, because I wanted very much to do my part and do my job well but didn't yet know how to do that properly. I was trying hard to take it all in.

That afternoon word went around that we were going to have a two-company perimeter that night. We were linking up with Alpha Company. We were told they would have half the perimeter and we would have the other half. We were linking up because there was a very good chance that a large NVA force was in the area, and we might be engaged with them shortly. As a precaution, battalion command wanted our two companies together to strengthen ourselves in the event of a future battle.

That news was very stressful to all the guys, but, I think, it was even more so to me, the rookie. Most of these guys had already been through one major battle with the NVA and knew what to expect. I only had my imagination to tell me what it would be like.

When we arrived at our night perimeter, my team leader told us we did not have to dig a foxhole. Charlie Company was supplying two of the four listening posts (LPs) that night and Alpha the other two. My team was going to be one of our company's LPs. As LPs, we would be positioned about 75 meters outside the perimeter all night, so we didn't need a foxhole. This was my first LP. I had no idea what to expect.

Just before dark, we got ready to go outside the perimeter. My team leader told me to just take my M-16, one extra magazine, and one grenade. We were given a radio and the four of us moved out to our overnight position, outside the perimeter.

As we moved out through the thick jungle, I kept imagining us trying to get back to the perimeter in the pitch dark. I had already learned how dark it was in the jungle during my previous two nights. I looked back toward the perimeter a few times on the way out but knew there was no way I was going to memorize any route back.

About 75 meters away from the perimeter, my team leader picked what he thought was a good spot for us to spend the night. It was between two fallen logs with some brush around us. We settled in.

My team leader told me that we were all supposed to stay awake and on watch all night, but we would not be doing that. He said that we all needed to get some sleep that night. So, we would be doing two up, two asleep, all night, rotating every two hours. That sounded very good to me because I was exhausted from the day's walk.

He also told me that every half hour one of the CO's RTOs would be calling us on the radio to request a situation report (sit-rep). Maintaining silence on LP was very important, so we didn't have to talk for the sit-rep if we didn't want to talk. We could simply respond with a single squelch if everything was OK, or with a double squelch if something was wrong. A squelch was produced by pressing the talk button on the radio and releasing it. Our call sign on this LP was 2-6 Echo.

Darkness came fast. The team leader and I had the second watch, so we both leaned up against one of the logs and tried to sleep. In spite of how tired I was, I found it hard to sleep very soundly. When our watch came, the two other team members woke us and said everything was quiet. Then they laid down against the log to catch a couple of hours of sleep.

In the pitch-black night, I didn't want to get too comfortable during my watch to prevent myself from drifting off to sleep. So, I just knelt on the ground with my M-16 lying next to me, with my left hand holding on to it. I held the grenade in my right hand. Time passed so slowly. Every half hour a call came on the radio for a sit-rep. My team leader answered with a single squelch.

Then, probably about 1:00 AM, a call came that was not a scheduled sit-rep. My squad leader quietly answered with a very soft whisper "2-6 Echo." He listened for a short while and whispered quietly, "Roger, out." Then he leaned over very close to my ear and whispered "One of the Alpha company LPs thinks they have movement. We are supposed to sit tight." He then quietly woke the two sleeping guys and whispered the same thing to them. I stayed in my kneeling position without moving a muscle.

The jungle around us was quiet. There were just some sounds

of insects, but overall, it was quiet. No movement near us. Time passed, but now even more slowly. Then another update came over the radio and our team leader passed it on to us. "They still have movement."

Now, it was getting very tense for everyone. I kept thinking of how to run back to the perimeter in the blackness that surrounded us if we had to. I felt I knew the general direction, but wondered how easy it might be to get turned around in the dark.

About another hour passed without any more updates. I was still kneeling on the ground, leaning on my arms, with my M-16 in my left hand and a grenade in my right. I didn't want to let go of either because, in the dark, I might not be able to find them again. I started to relax a bit. No news is good news, right?

I was half awake and half asleep, kneeling there when suddenly, without warning, in the blackness, a fully automatic burst from a weapon came from the direction of the Alpha company LP. I was fully awake in an instant. I had no idea what to do.

As I knelt there in the dark, I could feel myself start to shake. I thought I was shaking so bad that the whole perimeter must hear me. I immediately got very angry at myself. I tried to calm myself down and tried to stop shaking. I'm sure the shaking lasted less than a minute, but to me it seemed like it went on forever.

My team leader immediately got on the radio and asked what the hell was going on. Soon he was told that one of the Alpha LPs had taken it upon himself, without permission, to shoot at the movement he thought he heard. As of now, there was no follow up movement, so it was probably a false alarm. We should stay put.

I calmed down a bit, and the shaking had stopped, so I started to feel a little better. I was glad it was so dark. At least, my teammates hadn't seen the shaking happen. But, I wondered, had they heard it? What did they think of me? Coward? I didn't know. No one ever said anything to me. Maybe they were all shaking too?

The rest of the night passed very quietly and very slowly for me. I think we got back to our normal two on, two off sleep as best we could. Nothing came of the incident. The next day, Alpha and Charlie companies went our separate ways. No NVA were contacted in the area.

That night on LP was my third night in Vietnam. On that night, I experienced the most raw fear I would ever experience in Vietnam. From that night on, every other fearful situation that happened to me in Vietnam never compared to that first time on LP.

10. Land Leeches

Throughout the jungle areas of Vietnam are many environments that are very favorable for land leeches. Land leeches are small, maybe one to two inches long, black, slender, worm-like creatures. They travel across the jungle floor like inchworms. They can travel relatively fast for their size, especially if they sense a meal is nearby.

Normally, using lots of insect repellent on your boots and pant legs kept these little parasites from getting on you as you walked through the jungle. However, if you sat down at any time, you could pick up a traveling companion pretty easily. Once on your clothes, they crawled around on you till they either found bare skin or a way under your clothes.

If they got on your neck or face, you could usually feel them, or a buddy would see them, and you could knock them off before they started their lunch. If they got under your clothes, it was difficult to feel them. Also, once under your clothes, they could make themselves at home wherever they wanted because we very seldom took our clothes off. We only got a change of clothes ever week or two and the opportunity to take a bath or just wash up in a stream didn't come around very often.

We did do a self-check for leeches once in a while. We would take our shirts off and maybe pull down our pants and drawers and check ourselves all over, but that wasn't a daily thing. We were usually inspired to do it when we heard someone else had just discovered a leech somewhere on his body.

The leeches generally liked to find a tight place under your clothes to make into their traveling restaurant. So, under the belt line or in your belly button were the most common places guys found fat leeches curled up on their bodies.

Once found, we got them off by gentle pulling or pouring lots of salt on them. When they did come off, the wound they made with their mouth usually bled quite a long time. They had some kind of anti-coagulant in their saliva to help them get their fill of your blood.

The land leeches were incredibly tough. I remember trying to stomp on one the first time I saw a leech. I stomped down hard on it, and it didn't seem phased at all. I thought I must have just caught it between the cleats in the soles of my boots. I stomped again. Same result. That leech just kept coming. About the fourth stomp, with no apparent harm, I decided these leeches were indestructible and quit stomping.

To get our revenge on these little bloodsucking varmints, when we saw them on the ground, if we had time, we would spray them with insect repellent. That was fatal to them. They died horribly, which pleased us very much. As soon as the repellent hit them, they would straighten out and stretch themselves, pencil lead thin, to two or three times their normal length. Then they would curl up and die. Take that, sucker.

I was once walking along and clenched my fist for some reason and felt something cold and squishy under my thumb. I looked and there was a fat leech on a sore I had on my ring finger knuckle. Somehow, it had gotten on my hand and found the sore and started dining.

My immediate reaction to this violation of my body was to flick my ring finger as hard as I could. Off came the leech, flying into the undergrowth. Left behind was a bleeding knuckle. I just wrapped my finger in a handkerchief and soldiered on.

Speaking of sores, it was very common for small cuts on your body to heal very slowly. Cuts on the forearms or shins were very commonplace. Sometimes, they would just fester for days and days in the heat and humidity of the jungle. We called these unhealing sores, jungle rot. The worst was if you got sores on your feet. That made walking almost unbearable. I was lucky and never had foot sores happen to me, but friends of mine sometimes suffered a lot with them.

If the sores were on your hands, it was extra painful, too. To protect my hands from scratches as we traveled through the jungle, I had my folks send me a pair of work gloves. I wore them regularly. They worked.

11. Termite Columns

As we traveled through the jungle every day, you'd see one or more new insects you had never seen before. Lots and lots of strange bugs there. But there were also some insects you saw quite often. Bees, ants, and termites were the most common.

Termites were interesting and could provide a bit of entertainment as well. First of all, they were very impressive looking bugs. The termites in Vietnam were brown and were about the size of very large ants. Some members of the colony had heads about twice as large as their bodies. All had large, black mandibles that opened very wide.

Termites traveled through the jungle in long columns that were, maybe, six inches to a foot wide. Thousands of them were on the move. You saw neither where the column was going nor where it came from because it seemed to be endless.

If the column crossed your path, you jumped over it. No big deal. Sometimes it paralleled your path for a little while. Most of the time, when you saw it, you pretty much just ignored it. We had someplace to go today and so did they. They were just plodding along following the termites ahead of them, and we were just plodding along following the soldiers in front of us.

However, sometimes when our column stopped for some reason, and you just happened to be near a termite column, you were tempted to do a little mischief. This was especially true if the termite column was traveling down the length of the trunk of a large fallen tree. Often the column was just as wide as the top of the log. To mess with them a bit, as you watched the hundreds of termites pass by, you brought out your handy bottle of insect repellant. Then, in the middle of the log, you would spray the repellant across the diameter of the top, sides, and bottom of the log. The insect column would come to an immediate halt on one side of the repellent. On the other side, the column just continued on its way.

This was the start of a huge insect traffic jam. The termites near the repellent refused to cross that section of the log and all the termites behind them just kept coming. Soon that log was covered with termites wondering what the holdup was. We found it quite amusing doing that to the poor termites. We were easily entertained.

Unfortunately, just then, our column usually started up again and you had to leave. You never got to see the outcome of your little joke on the termites.

I confess to pulling this trick a couple of times.

However, the termites got even once.

We began using air mattresses just after the rainy season started. Until then, we just slept on the ground using our poncho liners as both ground cloth and blanket. But, when the rains started, we got tired of waking up after sleeping in the mud that formed when rainwater ran under our hootch. None of us wanted to carry the extra weight of an air mattress, but sleeping on one became a necessity if you wanted any chance at a decent night's sleep. So, we carried it.

Using an air mattress required some preparation. After laying the air mattress on the jungle floor, we'd spray all around it with insect repellent and also spray the edges of the mattress. This was our best way to try to keep the creepy crawlies from climbing up onto the air mattress with us as we slept.

One morning, I woke up and my air mattress was completely flat. No air in it at all. Confused, I pulled up the edge of the mattress to see if I had laid it on a sharp rock or something. When I pulled it up, underneath were hundreds of termites. Many of them were still biting at my rubber mattress. I must have laid my mattress on top of an exit to a ground nest. During the night they couldn't leave because of all the repellant sprayed around my mattress so they just began chewing on the mattress. End of mattress. Sweet revenge for the termites.

It took a couple of days to get a replacement mattress, so I had to, again, just make do in the jungles of Vietnam.

12. 'Snake' in the Weeds

One evening, dark had just fallen. We were all done building our night perimeter, had eaten our evening meal of C-rations, and had a little time to relax before going to sleep. I was sitting on the ground and quietly trying to fix some part of my equipment.

The area where I was sitting was covered with foot high weeds.

As I was concentrating on my task, I heard a very quiet "Shhhhhh" directly behind me in the weeds. I froze. To me, the only thing I could imagine making that sound was a snake moving through the weeds. The sound stopped for maybe five to ten seconds, then another quiet "Shhhhhh." Same place. The sound stopped again. After five to ten seconds or so, another "Shhhhh." Same place. No movement toward or away from me. Just right behind me. What the hell was it?

As this repeated itself for another minute or so, I began to doubt that it was a snake, but had not yet fully convince myself of that. I had a choice to make. Sit quietly and wait or make a move to get away from that spot. I chose to move.

I gathered myself and suddenly sprang up and away from

that area. Once I had a couple of steps between the sound and me, I relaxed a bit. I stood there and listened. Sure enough the "Shhhhhh" sound continued in the same way as before. Same place. No movement away from the area.

By then, I was convinced it couldn't be a snake making that noise as it crawled through the weeds. If it were a snake, after a couple of minutes of crawling, it should have moved completely out of the area. Couldn't be a snake.

I decided I had to find out what made that sound. I got a flashlight and quietly snuck up on the weeds. As I hovered over the spot that the noise was coming from, I brought the flashlight, still unlit, just over it. I put my fingers over the lens to eliminate almost all light from coming out and turned on the flashlight ready to leap back if necessary.

I saw no snake in the sliver of light coming out of the flashlight. I quickly scanned the light through the weeds nearest me and still saw no snake. Relieved, I began a more thorough search.

As the slender beam of light moved through the weeds, all I saw were ants. Lots of them crawling through the dense weeds. Nothing else. I searched for maybe five minutes and saw nothing else in those weeds but ants.

To this day, I don't know if I missed seeing something in those weeds, but I don't think I did. That evening, the only conclusion I could come to was that, somehow, those ants were making that "Shhhhhh" sound. I just don't know how. I don't remember ever encountering that sound again.

It became just one more of the mysteries of the Central Highlands of Vietnam.

13. 'Care Packages'

One of the things most appreciated by the troops was when we received a package from home. We called them 'Care Packages'. My folks (mostly my mom) were really good at regularly sending packages filled with goodies and requested items to me.

The guys always shared any edibles we got in the packages with our buddies. It was an unspoken rule to share. We usually received several days' worth of mail at the same mail call, so almost everyone got one or two packages and we had goodies galore for a couple of days.

Most times the packages arrived in decent shape with most of the contents relatively undamaged. However, sometimes the packages arrived very abused from their trip and only the toughest items inside survived. No matter what, one of the items that rarely arrived unscathed were cake items. They just didn't travel well. If someone received a birthday cake for instance, we usually just scooped it out of the package with a spoon because it was generally just a mass of squashed cake and icing. But, it still tasted good, and that was all that counted.

One of the things almost always in my packages was a small batch of my mom's chocolate chip cookies. Even though my dad ran a bakery, mom's home baked chocolate chip cookies were always our family's favorite cookies.

Another thing almost always included in my packages were a couple of rolls of instamatic film. I think I went through three of those small, easy to use Kodak Instamatic cameras during my tour. The heat and humidity and the jostling inflicted while traveling with an infantryman in Vietnam was often just too much for them. They usually only lasted a few months and then broke.

The pictures they took were OK, but not great. One short-coming was, they did not take very good pictures in low light. That was a real drawback because, in the triple canopy jungle we were usually in, low light was a way of life. But they worked ok, were inexpensive, and were small enough to easily carry in the jungle. I brought back 183 pictures from Vietnam that survived the camera and environment. But I wish I had taken three times as many pictures as I did.

Instant Kool Aid packets were always included in my pack-age too — at my request. It was the kind of Kool Aid where no sugar was needed. I requested a lot of them because we regularly added Kool Aid to the water in our canteens. Especially if that water came from some of the foul streams we walked through. You got your water wherever you could, and often bad stream water was your only choice. Iodine tablets killed the microor-ganisms but added to the bad taste. Kool Aid helped mask that taste.

Scratches to our skin had a hard time healing in the hot, humid jungle. They sometimes got infected and would fester and not heal for days. The best thing you could do for yourself was to avoid getting scratched. Normally I always traveled with my fatigue shirt sleeves worn down to my wrists, but my hands were still exposed. My solution to scratched hands was to ask my folks to send me some work gloves that I could wear as we traveled through the jungle. They sent them. The gloves worked well. I think I went through a couple pair of work gloves during my year. I imagine my folks were surprised the first time I wrote home and asked them to please send gloves to a tropical jungle in my next care package.

Sweets and other things to eat, besides the cookies, were usually included in the packages. I told my folks to stay away from sending chocolate treats — they always came melted and made a mess. Jiffy Pop Popcorn was popular. However, the only sources of cooking heat we had available were heat tablets and C-4 plastic explosive. Neither worked very well, but we learned that cooking Jiffy Pop Popcorn with C-4 almost always resulted in burned popcorn. C-4 just burned so darn hot.

I remember, one package I got contained a box of Bugles. I had never had Bugles before. The package also contained other edibles besides the Bugles. The other things were opened and the guys in the squad were all eating well. I was walking somewhere away from the guys when I decided to open the box of Bugles and give them a try. I tasted one and thought it was absolutely great. Just then, a good friend of mine walked by and I offered him a Bugle. He thought it was great too. I grabbed him by the arm and said come with me. We went behind one of the line bunkers and sat down out of sight of the others. He and I sat there and finished off the whole box by ourselves. I still feel kind of guilty that I didn't share any of that box with the other guys, but those Bugles were sooo good.

It was best if the care packages came when we were not going to be moving for a few days — like when we were guarding the firebase. However, if we got a lot of packages in the evening while we were out and on the move through the jungle, many times some of the food went to waste because we couldn't finish all of it (remember everyone shared their goodies—except Bugles—one time!). None of us wanted to carry extra weight, so we sometimes, reluctantly, left the uneaten goodies behind when we moved out in the morning. We usually buried them before leaving so the NVA couldn't enjoy them if they came in to examine our previous position after we left.

Care packages and mail of any sort were always very welcomed by us. They were our only connection to the folks, family, friends, and life we had back home. They always helped get our minds away, at least for a while, from the heat, humidity, and hard life we were living. "Mail call!" was one of the best sounds we heard in the jungle.

14. Mad Minutes

Charlie Company sometimes went for weeks without any contact with the enemy. During that time, most of us had no opportunity to fire our weapons. Not firing them meant that we could not be sure they would function correctly if we needed them in a combat situation. To remedy that, very occasionally the company had what we called a 'Mad Minute'.

The first mad minute I recall was scheduled for early morning, just before we moved out from our overnight perimeter.

After multiple checks to make sure there were no soldiers outside our perimeter, the word went through the company that the mad minute would happen soon. We all positioned ourselves near our foxholes along the perimeter and waited for the fire command.

When the command was given, over 100 soldiers began firing their weapons into the jungle beyond the edge of our perimeter at the same moment. M-16's, M-60 machine guns, M-79 grenade launchers, shotguns, and .45 pistols all went off simultaneously and fired almost continuously for about a minute. The guys with M-16's tested them on semi-automatic and fully automatic and could go through one or two magazines in that minute. It was an awesome sound, hearing so many weapons discharging at the same time. The firepower of an American Infantry company in Vietnam was incredible.

Once you found your weapon functioned correctly, if you wanted, you could quickly check to see if your sights were set accurately. You'd do that by aiming at a specific thing beyond the perimeter, shoot, and observe if you hit it or not. Checking your sights was easier if you were using magazines loaded with tracer rounds. It was easy to see where your rounds were going using tracers. Lots of guys using tracers also made the mad minute a visual wonder as well as an audio wonder.

Once the minute was up, the order to cease fire was given and the jungle turned silent again. Except, of course, for the ringing in our ears, and all the excited laughter of the guys who had just blown away so many leaves and trees just outside our perimeter. Everyone had a lot of fun during a mad minute. If there were enemy soldiers near enough to hear us, I can't imagine what they thought was going on.

15. Christmas Rest

It was December 1966. Our whole battalion was finally going back to the 4ID base camp for a 10-day Christmas rest.

I had joined the company 33 days before in the Central Highlands. Since then, we were out in the field continuously. We had probably only spent 10 or 12 of those days on Palace Guard. That was what we called it when we were the company providing security for the firebase. The firebase was still out in the jungle, but at least we didn't have to move every day and build a night perimeter every evening. The rest of those 33 days we spent humping through the jungles of the Central Highlands.

One of the bad things (among lots of others) about being out in the boonies was that you had few opportunities for decent personal hygiene. Showers and baths were out of the question. Teeth care was also difficult. I tried to brush my teeth at least every other day, but it only happened when I had the time and opportunity.

Washing up was also an optional activity, done only when time and water availability allowed it. Water was often hard to get, and you needed to drink a lot of it every day. Washing up was way down the priority list when it came to using water. I carried a small bar of soap, a washcloth, and half a towel for just those occasions.

About the only time we had enough water was when we happened to make our night perimeter near a stream. Then, if you had time after digging in, you could go down to the edge of the stream, get a helmet full of water and take it back to your bunker for a sponge bath from your steel pot.

Steel pots come in two parts, a plastic inner lining, and the steel outer portion. The plastic inner lining contained the webbing that goes directly on your head. The outer portion is the actual steel part of the helmet. It is just like a large pot that fits snugly over the inner lining. It was the outer part in which we carried the water and then used as our sink.

When you got back to your bunker with the water, you set your steel pot on the ground and ensured it wouldn't tip over. Then, using it as a sink, you washed, usually only your upper body. When you were finished, you rinsed out your washcloth, dumped the water, and felt good for a day or so. Unfortunately, you usually had to put your dirty clothes back on your clean body. We only got a change of clothes every week or two.

Because of the lack of hygiene, showers were our first order of business once we were back at the 4ID base camp. Right after we checked into our tent barracks and turned in all our weaponry, we gathered our soap, washcloth, and towel, and headed to the showers.

The showers were down the street from our barracks, and we all walked down there. The showers were just large tents with multiple shower heads set up in a long line. I took a long, hot shower trying to get 33 days of sweat and dirt off my body. The other guys had been out there much longer than I, so I'm sure it felt extra special to them.

When showers were done, we got clean clothes. The clean clothes were delivered to us in large laundry bags that were dumped on the ground. There were green fatigue shirts and pants and green undershirts, underwear, and socks as well. After the shower, we just picked through the piles to find sizes that were close to fitting. You could have any color you wanted, as long as it was green. After dressing, it felt great to be clean again. We just left our filthy clothes behind in a pile. I didn't envy the guys who ran the laundry and had to handle that pile of dirty clothes.

We were told that the Red Warriors were in complete stand down mode while in base camp, so we had no guard duty or any other responsibilities. Just rest.

One thing that was negative about base camp was that we had to play Army while we were there. That is, we had formations, we had chow schedules, we had to do police calls, and we had to salute officers. Much different than when we were out in the boonies. Out in the boonies, we were just 100 men supporting each other and trying to survive. None of the Micky Mouse Army stuff out there.

While we were in base camp, the Army supplied booze and beer for us, so we had nightly parties in our tent barracks (See Figure 15-1).

We also received a lot of our care packages when we got to base camp. That meant, we were all eating well, because we all shared all the food goodies that came in the packages from home.

On one package I got, my folks had marked 'Christmas Presents'. Our family tradition was always to exchange our presents on Christmas Eve, so I saved that package till Christmas morning. Vietnam was 13 hours ahead of the Midwest, so I waited till 9:00 AM to open my Christmas package. 9:00 AM Christmas morning in Vietnam was 8:00 PM Christmas Eve at home. Right when my family was opening their packages at home, I was opening mine in Vietnam (See Figure 15-2).

The Red Cross also provided us with some things we usually didn't get in the boonies. One of those things was a box full of paperback books, from which we could take any book we wanted. I had heard of a book called 'Catch 22', but never read it. When I saw a copy of 'Catch 22' in the box, I grabbed it. I read it during that Christmas rest period and thought it was the funniest book I had ever read. As I lay on my cot reading it, I laughed uncontrollably. All my buddies thought I was going nuts.

The highlight of the Christmas rest was the Bob Hope Christmas Show. I had grown up all my life hearing about Bob Hope. He was a great American patriot who loved the troops. Every Christmas he would assemble a group of entertainers — beautiful women, comedians, musicians, singers, and dancers — and, oh yes, did I mention, beautiful women. He traveled to the far corners of the world, taking his show to entertain GIs. He took the show to Vietnam every year the US was involved in the war. In 1966, part of his Christmas tour was a stop at the Fourth Infantry Division base camp.

About two hours before the show started, the 1/12th Red Warriors marched from our company area to the show ground. The show was held at the bottom of a large natural bowl in the middle of the 4ID base camp. Once there, we all settled into our seating area (grass on the ground) and waited for the show to start. Our area was about halfway back and just to the right of the stage. Not a bad spot, but I always thought we infantry guys should have been right up front. One good thing they did do was put a group of WIA guys right up front, so they had great seats.

Eventually the show started. Bob did his comedy routine and his female star for this ensemble, Joey Heatherton, was great. I already had a crush on her from seeing her on TV before I went into the Army, so seeing her in person, even from so far away, was great. The whole show was very memorable.

When the show was done, a few of us pushed our way closer to the front to get a better look at the entertainers who were still on the stage interacting with the GIs (See Figure 15-3).

Afterword, I was very glad that I had had an opportunity to see it. I felt seeing it gave me a closer connection to all those past GIs in WWII, Korea, and all the other areas who were also entertained by a Bob Hope Christmas Show.

In the 4ID base camp, latrine facilities were still very primitive. They were just large wooden boxes with three or four holes in them like outhouses with no privacy (See Figure 15-4). Under each hole was half of a 55-gallon drum. Periodically, the drums were dragged out, diesel fuel was poured in, and then set on fire. Guys doing that detail had long sticks that they used to stir the drum's contents periodically to ensure it all burned. That was a work detail, I'm very glad to say, I never had to do.

Since they collected all our weaponry when we got into base camp, the morning we were scheduled to leave, they brought out large boxes of grenades, ammo, trip flairs, claymore mines, etc. to the center of the company area. We were allowed to take as much of everything as we wanted before going back out to the field (See Figures 15-5 and 15-6).

We stayed in base camp, resting, for a few more days after

the Bob Hope show. Then we went back out to the boonies and resumed our lives in the jungles of the Central Highlands looking for the NVA.

Figure 15-1: Bill, second from the right, loved to sing and led a small group of us singing Christmas carols on, I believe, Christmas Eve. Just part of the activities as we enjoyed ourselves during our Christmas rest back at the 4ID base camp.

Singing Christmas Carols

Figure 15-2: When I opened my Christmas present from my folks, I purposely opened it at about the same time as they were opening their presents on Christmas Eve back home. Vietnam is 13 hours ahead of home, so I opened my present on Christmas morning.

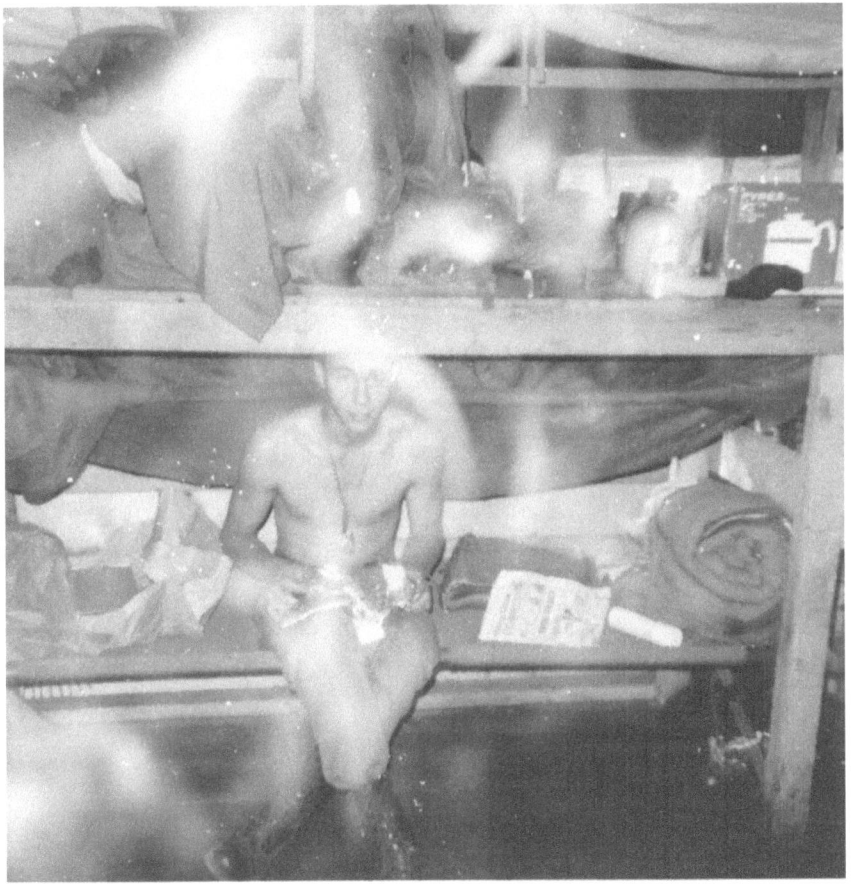

Opening Christmas Present

Figure 15-3: The show had just ended, and Bob and Joey Heatherton were talking with the crowd. Our seats for the show were significantly further back and, when the show ended, we moved up for a little bit better view.

Bob Hope Show

Figure 15-4: Primitive facilities, like outhouses without walls. At least reading material was usually left nearby to occupy your time.

Base Camp Latrine

Figure 15-5: We were unarmed while in base camp. The morning we were scheduled to go back out to the boonies, they brought out these boxes of weapons and ammo. Each of us could take as much as we felt we might need in the field. Most guys took all that and a little more.

Grenades and Ammo—Take All You Want

Figure 15-6: Staged up and just waiting for the call to head out to the choppers that would take us back out into the jungle.

Ready to Go Back Out

16. Typical Day in the Central Highlands

Most days we had no contact with the enemy. A typical day without contact in the Central Highlands for Charlie Company, 1/12th, 4ID on a search and destroy mission in 1966 / 1967 went like this.

We were awakened about an hour before dawn and immediately started to pack up as quietly as possible. Quiet was important because there was always a possibility that the NVA were near, and we didn't want to give them any advantage as to our location. Packing was almost always done in the dark. No flashlights were used for the same reason we tried to keep quiet.

Packing started by taking down your hooch which was made up of two ponchos snapped together and held up by a framework of, usually, bamboo sticks cut from the jungle nearby. The two ponchos belonged to the two guys who slept in the hooch.

I always kept my M-16 in the hooch with me as I slept. Also, when I slept, I never took my boots off. I untied and loosened them, but always kept them on. I did not want to find myself in an emergency situation in the middle of the night wearing just socks. For a pillow, I always used my rolled up fatigue shirt.

Putting my shirt back on in the morning took care of packing my pillow.

Our ponchos never served as raingear because we never wore them when it rained. When it started raining, we just walked on, getting soaked. Trying to stay dry was useless, even if we were wearing ponchos. Besides, ponchos were inconvenient, hot, and noisy to wear, so we didn't wear them. However, they did serve us in two ways, the first was as a hooch builder and the second as a protective wrap for stuff we wanted to try to keep dry. A third and fourth use, sad ones, were to carry a wounded Soldier to load on a Huey helicopter to evacuate to an aid station in the rear and to wrap the body of one of our Soldiers who had been killed. Fortunately, we had few of these last two uses.

Efficient packing was predicated by keeping all the gear you took out of your pack the evening before in a place where you could find it in the dark the next morning. That meant that despite being exhausted the previous evening, you still had to have the discipline to organize for morning packing before going to sleep. One never knew what the night or next morning would bring and it could mean your life if you couldn't find a critical item during an attack. Also, it might mean leaving behind a needed item if you couldn't find it while packing in the dark. So, a place for everything and everything in its place was my rule before going to sleep.

To start packing, after the hooch was down, we took our poncho and spread it out on the ground. Then we gathered up the items we wanted to keep dry. Things like our poncho liner (a lightweight nylon blanket), air mattress, stationary, letters from home, toiletry items, extra socks, and miscellaneous other small items, were put on top of the poncho and then the poncho was folded and rolled around them. These were always things we wouldn't need during the day's march or in a firefight. Once the poncho was rolled up, it was strapped onto our pack.

By the time we finished packing, it was just getting light. It was then we usually had time to fix and eat our breakfast of C-Rations. Everyone had their own tastes as to what they liked for breakfast, so we were all on our own. No mess hall or restaurant for us grunts out there in the jungle.

At first light, the company sent two or three sweeps outside the perimeter to make sure there were no enemy soldiers near our position. The sweeps were composed of four-man teams and each team had the responsibility of going out into the jungle a hundred meters or so and then circling the perimeter to confirm no enemy soldiers had snuck up close to us during the night. When the sweeps were going out, most of our soldiers were busy packing up or eating. The sweeps ensured that enemy soldiers were not close enough to us to surprise us and perhaps overrun us because we were distracted.

After breakfast, the officers and NCOs began to organize how we would move out. Each day we rotated which platoon (1st, 2nd, or 3rd) was the lead platoon (point platoon) and which was left or right flank. The weapons platoon was always the rear platoon. The command group was always in the center column of the formation.

Three- or four-man teams from each platoon were designated as the point element and the right and left flank security. The team from the lead platoon that day would be the point element. Our company moved in a formation of seven columns shaped somewhat like a diamond. The point element team was always 50 to 75 meters in front of the middle column. The flank security teams were 50 to 75 meters away from the main body of the formation as well.

In this diamond formation, the guys from each platoon were positioned in the same general area of the formation. If we had contact with the enemy and decided to quickly form a defensive perimeter, by using this diamond formation, all the guys from the same platoon would end up in the same section of the perimeter. That made for better command and control during a battle.

The point element was very important to our movement. Their first and primary job was to lead us in the direction we were supposed to go that day. Thus, they had to be good with compass and map. Their second job was to ensure that the whole company would not walk into an ambush. The third job was to cut a path through the thick jungle so the following soldiers could get through easier. Because some areas of the jungle required almost constant cutting to get through, the guys rotated who was first in line doing the cutting. The first couple of guys in each of the other columns and the flank security teams also had to cut their way through the undergrowth. Being the first couple of guys in a column always made for an even more difficult day of travel.

Once the order of travel was established, we got the order to "Saddle up!" and began to move out (See Figure 16-1). Amazingly, after doing it every day, getting ourselves into our diamond formation became fairly easy to do.

We always moved in a very spread-out formation. We tried to maintain a 10 meter gap between men in each column and, maybe, 20 — 30 meters between each column (See Figure 16-2)

Walking through the jungles of the Central Highlands was a big challenge. First of all, everyone was carrying a pack that weighed 70 to 90 pounds. The primary weight came from ammunition, fragmentation grenades, smoke grenades, claymore mines, trip flares, maybe a mortar round or a belt of machine gun ammo, and the stuff rolled in our ponchos. In addition to the packs, we all had our weapon, mine was an M-16, a five-pound steel pot, one to three days of C-Rations, three to four canteens of water, boots, and clothes. As an RTO, I also carried 27 pounds of radio, batteries, and antennas. When added up, my pack was almost 90 pounds, and my other gear was about 25 pounds.

Adding to the difficulty of traveling through thick jungle that you had to cut your way through, was the terrain itself. There were the hills, valleys, and waterways we had to ascend, descend, or cross.

Each day we had an objective we were supposed to reach. When it was possible, we tried to have the center column of our formation walk the ridgelines that led in the general direction of our objective.

If you were in the center column walking along a ridgeline, the pace was fairly slow, and the ups and downs of the terrain were normally not too extreme. The guys in the flanking columns, walking on the sides of the hills, usually had more difficulty.

The terrain there was never kind to you. The flanks of the tall hills we were in were always composed of a series of undulating gullies and ridges caused by water erosion. You had to make your way down into the gully and then work your way up the next little ridge. Ahead, another gully and ridge awaited you. It was almost always much more difficult and much slower going for the flanks than for the guys in the center column.

The center column constantly had the accordion effect that happens to a long line of people walking through difficult terrain—slow down, stop, jog to catch up, stop, slow down, jog, etc. The accordion effect was much more pronounced for the columns on the flanks. That was especially true if the company leaders in the center column didn't think about the more difficult terrain for the flanks as they moved along the ridgeline. Good leaders made sure that the flanking columns were keeping up with the center column and that we were always in a proper battle formation.

Sometimes our objective was on the other side of a steep hill from where we were. If so, we had to walk/climb up the side of the steep hill and once at the top, walk/climb down the other side. Some of those mini mountains would take us hours to ascend and descend.

As we walked through the jungle, we had to constantly stay alert for the enemy. That was especially important for the point and flank security guys. If you were in one of the interior columns, you had the luxury of not needing to be quite as alert.

If you were interested, you could spend a little time observing the jungle through which you were walking. There was always a new bug that you had never seen before or some weird plant or flower. Sometimes, not often, there were the sounds of animals near us. Occasionally, a flock of parrots or a troop of monkeys were in the trees above. But, for the most part, I think a hundred soldiers moving through their jungle kept most of the wildlife at a distance.

As we moved through the Central Highlands, we regularly came across a watercourse of some kind. Sometimes it was a little brook you could step or jump over, sometimes a shallow creek filled with rocks, sometimes a knee-deep stream, and on rare occasions, it was a chest deep river. We crossed them all by just walking across through the water. Our boots and feet and parts of our clothes were wet most of the time.

When the column stopped moving for some reason (proba-bly the accordion effect), I usually just stood and waited for it to start again. I rarely sat down to rest unless I knew the stop was going to be for at least several minutes. When I sat down, it was a real struggle to get back up because my pack was so heavy, so I usually choose to just stand and wait.

However, there was one trick I had to help me rest better while I stood there. I'd curl up my toes in my right boot, lean over, place the barrel of my M-16 on top of the empty front part of my boot, and rest on the butt of my rifle, like a crutch. Doing that helped take some of the weight of the pack off my back. The weapon was on safe, and I didn't have any toes under the barrel, so I convinced myself it was not real dangerous to do that.

When resting like that, I often took my helmet off and laid it on the ground in front of me. There, inside of the webbing of the helmet, looking up at me, was a carefully folded Playboy centerfold. She was always nice company for a tired soldier on the move.

When we stopped moving for a couple of minutes, if I studied the jungle floor around me, sometimes I would see small black land leeches inch worming their way toward me from multiple directions. They must have had some heat or smell sense that told them a potential meal was in that direction just waiting for their hungry little jaws. When they got to me, they were disap-pointed. I always sprayed my boots and pants legs with lots of insect repellent. Leeches hated that stuff. They would get to the boot, try to latch on and then instantly pull away. They usually made multiple attempts on different parts of the boot but failed every time. Thank goodness for insect repellent.

There were also other natural hazards of the jungle to contend with—things like bees, fire ants, snakes, bamboo thickets, and wait-a-minute vines. Bees, fire ants, and snakes could hurt you.

The bamboo thickets sometimes made you lie down and do the low crawl just to get through them. Or they forced you to try to squeeze through spots between bamboo branches too narrow to squeeze through. Very frustrating.

Wait-a-minute vines were soft vines that had small sharp needles that were bent backward. If you pushed your way through a thicket that contained a wait-a-minute vine, the vine caught your clothes and/or skin and held on. When that happened, you immediately stopped and said, "Wait a minute" to the guy behind you as you struggled to unleash yourself from the clutches of that damn vine. The worst was when the vine's needles were stuck in your hands, nose, or cheek.

While we were on the move, sometimes you had to answer a 'call of nature'. If it was number one, you simply waited for your column to come to a brief halt, which happened often, and picked a nearby bush to use as your bathroom.

Number two was a different story. While the company was moving, you couldn't just make everyone stop while you did your thing (unless you were the CO, of course). You had to time your bathroom break for when you knew the company was stopping for a couple of minutes at least. But that information wasn't always available to you. So, sometimes, especially if it was an 'emergency', you just had to take a chance.

When you made your decision to go for it, you quickly took off your pack, grabbed your entrenching tool, and headed into the bushes nearby. Once in the bushes, you quickly dug a shallow hole, a cat hole we called it, and did your deed in the hole. They encouraged us to use cat holes to hide traces of our passage through the area as much as possible. I don't know if it did much good, but I always used a cat hole like they recommended.

After finishing up using the TP that came in the C-Ration meals, you covered everything up with dirt, and quickly got back to where you left your pack. Hopefully the column hadn't started yet so you could get your pack back on and get going without skipping a beat when the column did start up.

Sometimes, the column would start before you were finished. If it did, when you got back to your pack and put it back on, you could either hustle to catch up with where you were originally in the column or, most often, you just took your place in the column where you rejoined it.

If number two called while you were in the night perimeter, you took your entrenching tool and went just outside the perimeter to do your job. I always notified the guys at the foxhole I was near that I was going out there. I didn't want them to be surprised, and accidentally shoot me as I was coming back into the perimeter.

There is one positive thing concerning the above 'bathroom' activities. Using a cat hole was a skill I utilized for the rest of my life. I used it whenever I was out on a long hunt in the woods or on a wilderness fishing trip. Proof that the Army does teach you some useful skills.

Around mid-day, the CO would usually call for a rest and lunch break. We would all stop in place. I sat down and took my pack off for lunch. Naturally, lunch consisted of C-Rations. Most times we didn't have time to heat them for lunch, so we just opened a can of what we felt tasted the best cold and began to eat right out of the can. Most meals of C-Rations were eaten right out of the can, cold or hot.

A few cookies or crackers often served as my desert. Lunch

was also when I usually ate one of my cans of fruit. The different fruits in the C-Rations were my favorite thing to eat. They tasted very good, and the juice was always a great thirst quencher. It was a wonderful alternative to our iodine flavored stream water.

When we crossed streams, I almost always took the opportunity to fill my canteens. I put one or two iodine tablets into the canteen after filling it to kill off any microorganisms. But the iodine gave the water a bad taste. To help with that, I had my folks send me packages of pre-sweetened Kool-Aid to pour into my canteens to improve the taste of the water. It worked pretty well.

After lunch and a bit of a rest, we'd continue our march. Sometimes our mission was just to pass through an area looking for the NVA or for any signs of NVA activity. Other times the battalion commander gave us a mission to get to a specific location to check it out. Either way, we'd normally stop our movement by mid-afternoon so we had time to dig in before dark.

An hour or so before he wanted us to stop for the day, the CO would radio to the point element and tell them to start looking for a good spot for our night perimeter. A good spot was normally high ground—the top of a ridge or hill for instance—that was big enough to fit our normal size perimeter. Our perimeter consisted of about 20 foxholes, built about 10 meters apart, forming a circle. Those foxholes were built and occupied by the members of 1st, 2nd, and 3rd platoons.

The command group and weapons platoon normally built their foxholes in a smaller perimeter inside the main perimeter. The weapons platoon also set up emplacements inside the perimeter for the 81mm mortars they carried. The reason for a smaller perimeter inside the main perimeter was to have a second line of defense in case our outer perimeter was overrun.

Once a location for the night perimeter was found and the officers and NCOs designated where the foxholes should be built, we all started working hard to get it done.

Most guys did the same job(s) every night. One guy might like to dig the foxhole, another guy might like to set up the hooches, another might cut trees for overhead cover for the foxhole, while another might like cutting our field of fire (a cleared out area in front of each foxhole). We also had to put out our own booby traps for the enemy — trip flairs and Claymore mines. We all had our favorite jobs. Several guys from each platoon were always detailed to help cut the LZ.

While cutting the LZ, many times some of the larger trees were just too large to cut with a machete or an ax. In those cases, the team of engineers that were attached to our company were called on to bring down the large trees with C-4 plastic explosive (See Figure 16-3). The engineers would pack the base of the tree with the C-4 and insert the blasting cap into the C-4. Just before setting it off, the engineer would loudly yell, "Fire in the hole!" When you heard that yell, if you were close to the LZ, you quickly found cover. The blast sometimes sent wood and rocks out in every direction. The big tree that was being blasted would jump up in the air at the blast and then slowly topple over.

My specialty was cutting overhead cover. Once we stopped moving, I'd take off my pack, set it down near where our foxhole was going to be, grab my M-16, an extra magazine, my razor-sharp machete, and head outside the perimeter. Out there I'd look for a long trunked tree about eight to ten inches in diameter and cut it down. Once down, I'd cut the trunk into logs eight to ten feet long and carry the logs back to our foxhole (See Figure 16-4).

By the time I had enough logs, maybe six to eight, the guy digging the foxhole would be done too. We'd then fill a dozen or so sandbags and pile them up on either side of the foxhole. Once the sandbags were in place, we'd lay the logs on the sandbags over the top of the foxhole, the long way, giving us some overhead cover.

The purpose of the overhead cover was to provide protection in case the bad guys started dropping mortar rounds on us. The hope was, if a mortar round was going to land in our foxhole with us, it would, instead, hit the logs above us first and explode without injuring us. That was the hope anyway. Once we had our foxholes done, we sometimes took time to eat before continuing the rest of the work.

When the LZ was finished, helicopters came in with our supplies (See Figure 16-5). Usually we got C-Rations, water (if needed), and 4s and 5s (beer and soda). Sometimes mail was also included (See Figure 16-6).

When all the jobs were done, it was usually close to sundown. If we hadn't eaten before, we ate then. Suppers were almost always just heated C-Rations.

Just before dark, three four-man teams went outside the perimeter about 75 meters to serve as our overnight listening posts (LPs). Their job was to be our early warning system in case the enemy tried to sneak up on us during the night. They usually picked a spot that hid them well. At least two members of each team were supposed to be awake throughout the night. Each team had a radio they used to communicate with the company command group.

If the LP heard the enemy in the dark, first they radioed in the information, then they were given instructions from the Company Commander. Depending on the situation, they might be told to sit tight and observe for a while, or they might be told to throw a grenade or open fire on the enemy and then hustle back to the perimeter (75 meters) through the dark jungle as fast as they could. As they approached the perimeter, they were supposed to be yelling "LPs! LPs!" so the guys in the foxholes didn't shoot them as they came in.

When I was in second platoon, I was on LP a few times. When I was the CO's RTO, I had radio watch nightly, monitoring the LPs. We never had any actual enemy contact in the middle of the night during my time with Charlie Company, but there were some memorable false alarms.

Meanwhile, back on the perimeter, just before dark, the four men sharing each foxhole determined their watch sequence. We divided the dark hours into four 2+ hour shifts. Each man took a turn. One man from each foxhole had to be awake and on watch throughout the night.

During your watch, you simply sat on or near the foxhole and pretty much just listened. If the bad guys made it past the LPs, it was your job to discover them by sound alone. It was usually so dark in the triple canopy jungle you couldn't even use your time while on watch to write letters. After an exhausting day, staying awake during your watch was sometimes the most difficult job you had as a soldier.

If it was raining, you just sat there in the rain. The only cover over your head was your steel pot. Heavy rain made me nervous because the sound of the rain could cover the sound of any enemy soldier sneaking up. That was the time to be extra alert.

As I previously said, during my year a middle of the night sneak attack never happened to my unit. I think it was because the enemy was as affected by the pitch darkness as much as we were. It is very difficult to move through a thick, pitch-black jungle quietly and efficiently.

For most of my time with Charlie Company, we slept in what we called hooches. Hooches were built using two ponchos and several straight sticks (preferably bamboo). For much of the time, we used air mattresses to sleep on.

We always carried and used mosquito netting inside the hooch. Vietnam has a very high incidence of malaria, which is carried by mosquitos. The netting was used to try to reduce the number of guys who caught malaria. In addition to the netting, each of us was supposed to take tablets every day which also helped prevent getting malaria. I was very conscientious in taking those tablets and always used mosquito netting. I still came down with malaria a year after I got back to the States. Oh well.

About an hour before dawn, the soldier on the last watch woke his foxhole buddies and a new day started that was going to be pretty much the same as the previous day.

Figure 16-1: This is Charlie Company getting organized to move out of a firebase. Despite the apparent chaos, we got pretty good at getting ourselves into the diamond formation we used to move through the jungle every day.

Getting Ready to Move Out

Figure 16-2: A picture of a few soldiers from Charlie Company as we moved through the jungle. Notice the distance between them. This shows a column that was moving along a stream bed. I am in the adjacent column moving along a hillside. Note the bamboo between me and the guys in the stream bed. Bamboo thickets were everywhere in the Central Highlands.

Keep Spread Out

Figure 16-3: This helicopter was able to actually touch down in this LZ. Whenever they came in, several guys were waiting to go out to unload it. We didn't want the chopper to be in the LZ too long. Choppers were primary targets for the bad guys, if they were in the area. Note the splintered tree in the foreground. It was apparently brought down using C-4. C-4 was usually only used for big trees. We must have been in a hurry to build this LZ, so C-4 was used to speed up the job.

Chopper in the LZ

Figure 16-4: This is me doing my 'regular' job of cutting overhead cover for building the night perimeter. It took six to eight logs like this to properly cover a foxhole. This photo was taken in the Ia Drang Valley. The Ia Drang was where the battle was fought that was the basis for the movie "We Were Soldiers". That battle occurred here only one year before this picture was taken. The Ia Drang was also the place where the Red Warriors had a very costly battle in July of 1967.

Overhead Cover Freshly Cut

Figure 16-5: Whenever a helicopter came into a jungle LZ, a soldier would help guide it in to either a safe landing spot or to a spot where the helicopter could safely hover. Many times, the pilots didn't actually land because of all the trees and branches on the ground. Note the residual smoke in the air. This was smoke from a smoke grenade that was used to help the pilot identify our location in the jungle.

Guiding a Chopper

Figure 16-6: Whenever we got resupplied, the supplies were brought to a single spot within the perimeter. On this day we got C-Rations, Beer and Soda, and our mail (the two bags). Once there, it was inventoried and then distributed to the men. We might get one to three days of C-ration meals delivered. We all had our favorite C-Ration meals. There were also meals we didn't like—at all. When it came time to distribute three to nine meals to each man, the NCO's would open the cases of C-Rations upside down so the meal names were not visible. Then we would file past, taking our meal boxes randomly out of the cases. That way, the first guys through didn't necessarily get all the best meals. As soon as the picking was done, we'd turn over our meal boxes to see what we got. Every once in a while, you'd hear "SOB, Ham and Lima Beans again."

C-Rations Delivered

17. Carrying LAW

Every so often, the Army saw fit to send out a new weapon or piece of equipment for us grunts to try out to see if they would add to our effectiveness as a fighting force. Once, they sent out a Starlight Scope which was an early night vision device. It worked ok, but it was as big as a large telescope and came packed in a large, black, plastic suitcase. A few days after it arrived, our first sergeant sent it back. Too bulky, too delicate, and much too difficult to carry a suitcase through the jungles of the Central Highlands.

Another new piece of equipment they sent out for us to try was a Light Anti-Tank Weapon (LAW). They said if we ever encountered an NVA bunker complex, we could use it to attack the bunkers. I was the unlucky guy they selected to carry it all day.

A LAW was a one-shot self-contained bazooka. It weighed about six pounds and was a tube about 25 inches long. It had a sighting system mounted on the outside of the tube. It came with a sling for carrying. The way I settled on carrying it was to put all my gear on first, then sling it cross wise over my body with the LAW diagonal over my backpack.

Now, six pounds wasn't a lot of extra weight so that was not the real problem carrying a LAW. The problem was, it reached out and grabbed every vine and branch I walked by. I don't know how it did it, but I was constantly stopped in my tracks as I walked through the jungle when it grabbed a vine and hung on. And it was very stubborn when it got hold of a vine. It just didn't want to let go. So, there I'd be, pulling like a draft horse trying to break free of the vine, but not always having the strength to do it. Sometimes it would break free unexpectedly and I was lucky not to lurch forward and fall on my face. Since it was on my back, I couldn't see it to determine which way I should turn to loosen the vine's grip. If I turned left, that was wrong. If I turned right, that was wrong. Very frustrating.

Bamboo stands were another major source of irritation. Sometimes we would have to bend over to get through the arching bamboo stalks. Sometimes we would have to crawl to get under the bamboo. Other times we'd have to squeeze through narrow gaps in the bamboo to get through. Carrying a LAW on my back made every one of these maneuvers much more difficult. That LAW liked to grab on to the bamboos stalks as much as it liked vines.

After about a week of carrying the LAW, my platoon sergeant asked me how it was going. I described the problems I was having with it. Thankfully, a day or so later, he took it away and sent it back to base camp. His reasoning was that we rarely encountered a bunker system where it maybe could be useful and we were never going to encounter a tank, so why put up with carrying it? I loved him for that.

18. Ambush

It was early in 1967. I was a rifleman in Charlie Company, 1/12th, 4ID serving near the Central Highlands of Vietnam. Our company was operating in an area that had some civilian villages nearby. This was unusual for us because we usually operated far out in the jungles where there were no civilians anywhere near us. Out there, it was just us and the NVA. Out there, the rules of engagement were very simple, if you encounter anyone who is not a GI, shoot him. It was much less complicated than when civilians were in the area like they were now.

We were told that some of the villages near us were suspected of having VC and NVA sympathizers living in them. Because of that, the Vietnamese government had given the villagers a strict curfew to abide by. Charlie Company was given the job to help enforce the curfew. To do that, we established a patrol base that was probably one or two klicks from one of the villages.

One of the ways the curfew was enforced was to have random ambushes set up in the area to take out any VC units operating under cover of night. The area was quite heavily forested but had a network of foot trails everywhere.

My rifle squad was selected to set up an ambush outside a village along one of the main trails leading into it. We were told that no one was permitted outside the village after sundown. If anyone passed by our ambush site after that time, they were assumed to be VC, and we were ordered to open fire on them.

In early afternoon my eight-man squad left the patrol base and took a very circuitous route to our planned ambush site, which was several hundred meters outside the village. When we quietly arrived at the site, our squad leader positioned four of us on one side of the trail and four of us on the other side of the trail. We were well concealed in the thick undergrowth of the jungle.

Before we took up our positions, the squad leader told us what time curfew started. He then very sternly said that anyone walking the trail after that time was VC and could be shot. But, he also very sternly said that we were to hold our fire, until he shot first. No one was to shoot until he shoots!

I was with three guys on one side of the trail and the squad leader was with the three other guys on the opposite side of the trail. There was no way to communicate with the other group without compromising our position. Silence and stealth were imperative to our survival.

Time passed very slowly as we lay there, hidden in the undergrowth, waiting to see what would happen. I think this was the first ambush for all of us, so that added greatly to the tension.

As the light started to fade, the curfew time slowly approached. At that point, no one had yet used the trail we watched. The moment curfew time arrived, we all gave each other a tense look. I quietly took a deep breath.

About ten minutes later, we began to hear voices approaching our position on the trail. Three men were coming, heading toward the village. All were talking loudly and laughing a lot. Each of us quietly adjusted our positions, readied our weapons, and waited very nervously.

As the men passed the spot where we had planned to begin firing, the squad leader withheld his fire. All of us followed his lead and held our fire as well. The three men, all of whom appeared to be unarmed, passed by our position still talking and laughing, with no harm coming to them. They never knew they came so very close to death.

We stayed in our positions for the rest of the night. No one else came down that trail. At dawn, we made the long walk back to the patrol base. Our squad leader told us he withheld fire because it was just barely past the curfew time, and anyone can run a little late. Also, they were unarmed and all the talking and laughing they were doing made it sound as if they had a little alcohol in their systems. He felt if they were VC, that was not how they would have acted.

I agreed with him entirely.

I will never know if those men were VC or not. But I do know that I am forever grateful I did not have to wrestle with that question for the rest of my life. I think having the picture of three, possibly innocent, dead men in my mind would have been very hard to live with.

I'm very glad that my company didn't work very much in areas with a lot of civilians. Because of that, I was never involved in any actions that involved civilian casualties. I was very fortunate to come home from Vietnam with no pangs of guilt for any actions where civilians were hurt or killed. Just so lucky.

19. Bees in the Stream

As we moved through the jungle, we came to a wide shallow stream (See Figure 19-1). As I entered the shin deep water, I looked left and right. To my right I saw another soldier entering the stream at the same time. It was an NCO from my platoon.

Because we always traveled through the jungle well spread out, he was about 25 to 30 meters away. Being spread out was a safeguard against having a high number of casualties if we were ambushed. Less guys congregated together meant less targets for the bad guys, which, in turn, meant fewer casualties.

This sergeant always traveled with a green towel around his neck to help him wipe off sweat. When I saw him, I noticed he didn't look like he normally did. He had his towel draped over his helmet and he was slapping at the air around his head. He was also walking faster than normal through the stream.

It only took me a moment to realize that the sergeant was being attacked by bees. The bees in the jungles of Vietnam were very aggressive. If a soldier walked by their hive and somehow got them angry, they would attack any moving body in the area. Apparently, the sergeant was either the cause or an innocent bystander of the bee attack. It didn't matter to the bees.

As I watched him hurrying across the stream, I started feeling sorry for him. But, at that very moment, I saw and heard a bee come flying and buzzing, at high speed, straight at my face. It hit me in the middle of the forehead and stung me immediately. Now I was the one being attacked by bees.

I quickly broke into a very fast but measured pace going across that stream. I was trying to get away from the bees, but also did not want to fall down in the stream. I thought, if I fall, they will get me good!

Surprisingly, once I reached the edge of the jungle on the other side of the stream, no bees followed me, and I got away with only one bee sting. However, the way that one bee attacked me reminded me of a bee attack in a cartoon. In a cartoon, the bee flies straight at his target with his stinger leading the way. It seemed to me that is exactly what this bee did.

Figure 19-1: Crossing streams was almost a daily occurrence. We'd just walk across. Wet boots and pants were always a way of life. This stream looks like a good one for refilling canteens. The water looks decent. Most were. I usually filled my canteens no matter what it looked like. The soldier crossing looks hot and tired to me. We probably had been walking for several hours already when I took this picture.

Typical Stream

20. Night March to Village

We were working in an area that was mostly sparsely treed grassland. It was one of those rare areas in Vietnam where US tanks could operate. For a couple of weeks, we were assigned to work with tank units on several different missions.

Charlie Company was in a patrol base. One morning we received orders that we had a mission that required two platoons. I was in 2nd platoon, and we were one of the two platoons selected to do the mission. We were told to try to get some sleep in the afternoon because we were going to start our mission that night after dark.

Late in the afternoon, the mission was described to us. We were told to take only our weapon, ammo, grenades, and water with us. No other gear. Our mission would start after nightfall. We were to walk to a Montagnard village that was a couple of miles away, surround it in the dark, and wait till dawn. Tanks would come to reinforce us at dawn. The village was suspected to be a VC stronghold so we would be searching it for weapons caches.

After organizing our gear, we waited till dark. When full darkness fell, a local Vietnamese soldier or government official came to our patrol base. Serving as our guide, he led the two platoons out of the perimeter.

It was particularly dark that night. No moon at all. To prevent us from getting separated in the dark, we did the unthinkable, the two platoons walked single file down trails fairly close together. Never did we operate like that. Never on trails, never bunched up. But the circumstances that night made us change from our normal operating tactics.

There was a section of the route to the village that passed through a thickly wooded area where it was absolutely black. The only way any of us could keep in contact with the man ahead was to hold on to his belt. You couldn't see him or the terrain at all. As you walked behind him, like a blind man, the only way you could tell if the trail was going up or down was to feel his belt go up or down. It was very challenging walking that way.

Eventually we broke out of the pitch-black trees into a grassy area where at least there was some residual light from the stars. We started walking in two columns down a wide trail that was like a small road. As we silently walked, we suddenly got word from behind us that voices were heard. We all quickly and quietly moved into the long grass on either side of the trail and laid down.

Soon, two men came walking down the now empty trail. They were talking to each other, but not too quietly. Suddenly both platoons stood up and closed in on them from the darkness. They stopped talking immediately and got very wide eyed. The Vietnamese guide immediately had them tied up and began quietly questioning them. He questioned them for a few minutes and then told our officers to bring them with us.

It was probably 1:00 or 2:00 in the morning when we got to the village. We quietly encircled it, about 75 meters away from the edge of the houses. From what we could see in the darkness, the village appeared to have 10 to 15 thatched houses built on stilts. It was a Montagnard village. We formed ourselves into teams and took up our positions around the village to wait for dawn. We all hid ourselves in the long grass.

At first light, just before dawn, I could see that my team was positioned on the far side of a large garden area, behind the village. Just after dawn, an old man came out into the garden area. He was dressed in what looked like a nightshirt. He had no idea we were in the long grass just beyond the garden. The old man looked around a bit in the garden, then squatted down and did his business between the rows of plants. When he was done, he stood up and walked back to the huts. We just looked at each other and shook our heads.

Shortly after the old man left, we began to hear the sounds of tank engines in the distance. Soon, five tanks approached the village from five different directions. They quickly positioned themselves between the teams of soldiers on the outside of the village with their cannons pointed into the village.

When the tanks arrived, the villagers, having heard them, began to come out of their huts. They stood in an open area about in the middle of the village looking at the tanks surrounding their village. When the order came for us to stand up, we all stood and showed ourselves. The villagers looked even more concerned.

We all moved into the village with the tanks staying in place just outside it. The Vietnamese official with us took charge. He told our officers to have us search all the houses in the village. We should look for weapons or anything unusual. My team was told to search a particularly large house near the open area of the village. Other teams searched the other houses.

We searched the house as best we knew how. None of us had ever done this before. We looked for any secret spaces or trap doors where a weapons cache could be hidden. We found nothing.

I felt very uncomfortable doing that search. I knew it was a war situation, but it still felt to me like a very big invasion of privacy. Maybe I'd have felt differently if some of our soldiers had been hurt recently and we suspected these villagers had something to do with it. Don't know.

Once the searches were done, the infantrymen had nothing more to do, so all we did is stand around watching the interrogations of the villagers. The interrogator sometimes got a little rough with the guy he was questioning, but nothing too bad. I don't know if he learned anything or not.

One other thing that kept our minds occupied were the young ladies of the village. Montagnard women, most of the time, do not wear tops and, therefore, their breasts are exposed. Normally this was not a big deal because American soldiers usually did not find most Montagnard women particularly attractive.

However, it is my opinion that this village must have been near a French garrison 17 to 18 years before, because there were several exceptionally pretty, young ladies in the village. All without tops. To me, they did not appear to be full blooded Montagnards. I was, and still am, very sorry I left my camera behind at the patrol base the night before.

When we got back to the patrol base, we had all kinds of stories to tell the guys who didn't go on the mission. As I think back on it, that mission was more like a WWII war movie than anything else I did in Vietnam.

21. Tanks vs Nature

We were operating as infantry support for a tank unit (See Figure 21-1). We only did this for a couple of weeks during my tour. The area we were in was primarily grasslands and brush, with dirt roads scattered throughout the area. Unlike the Central Highlands, where we normally operated, this area was suited for tank operations.

Our mission that day was to move through a large roadless area of grass and brush, intermixed with the tanks, searching for any VC activity.

As we moved through the area, the tanks simply plowed through the grass and brush. We followed closely behind the tanks. I don't remember how many tanks were intermixed with our company.

The mission was almost over. We had just about reached a road that had trucks waiting to take us back to our perimeter when one of the tanks came to a fast stop. The hatches in the top of the tank flew open, and the five or six crewmen scrambled out of the hatches as fast as they could and jumped off. The tank was running, but not moving when they bailed. We infantrymen near the tank stopped as well, wondering what was going on.

When the tank crew hit the ground, they all started to do a strange dance and began to rip off their shirts. All of us infantrymen immediately knew, from our own experience, exactly what was happening to them. They were under attack from fire ants.

Fire ants made their nests in the leaves of bushes. Apparently, one of the bushes the tank ran over contained a fire ant nest. The ants, deprived of their home, became very angry and invaded the machine that had attacked them. The tank crew were fair game to the ticked off ants.

As a kid, I always liked to watch ants. The ants in the States that I was familiar with had mandibles that opened up like a 'C'. The ants in Vietnam had mandibles that opened up like a '['. Their bite was much harder than the ants back home. The tank crew was learning that right then.

All the infantry guys watched the dancing tankers with smiles till we had to leave. I never found out how that particular little battle turned out and have often wondered how that tank crew recaptured their own tank. But, knowing those ants, I'll bet they were still exacting their revenge with surprise attacks from within the tank days later.

Figure 21-1: Riding on a tank was a welcome change to slogging through the jungle, although a very dusty change. This picture of me was taken by a good friend of mine who, sadly, was killed a few months later by small arms fire.

Providing Tank Security

22. Tank Fire

We had just come back into our patrol base after a long day providing infantry support for the tanks searching for VC activity in the area. It was late afternoon, just a little before sundown. Our perimeter was on a grassy hill, and we could see a couple of miles away over the grasslands.

My team just got back to our foxhole area when someone said, "Look out there." Off in the distance, maybe half a mile away, a small grass fire was burning. Near to it was one of our tanks.

As we watched, the tank began driving over the burning grass, apparently trying to put it out. Suddenly, it stopped, and the tank crew began to clamber out. We could see that the tank's tracks on the side facing us had caught fire. The crew began frantically trying to put the tank fire out.

But it seemed the harder they worked, the higher the flames went up the side of that tank. Soon, it seemed they decided that it was useless trying to put it out, and they quickly abandoned the tank and the area. When they did that, we all settled in to watch what turned out to be a pretty good show (See Figure 22-1).

That tank was ready for combat, so it was full of munitions. That's obviously why the crew decided to put a lot of distance between themselves and the burning tank. They were right to do so.

Just as the sun was setting in the west, with the tank in the foreground, the explosions began. We all had expected the tank to have one grand explosion like in the movies. This tank had a different idea. It began to explode pretty much one round at a time.

Some guys were giving a running commentary about the event we were watching. "That was an HE." "Oh, wow, that was a Willie Peter." "There goes the machine gun ammo." It was actually good entertainment for a bunch of weary soldiers.

When the explosions finally seemed to be done, there was a grand finale. It came in the form of the round left in the chamber of the tank's cannon. When the tank was abandoned, its muzzle happened to be pointed to our right. Lucky for us it was not pointed in our direction. When the round in the chamber finally went off, we could hear it whistling down range to some unknown destination. We all hoped it landed harmlessly somewhere.

The tank burned well into the night. While it did, it was a good show for all of us grunts.

Someone said that he thought that a tank cost about $250,000. A significant amount back then because a decent new car went for about $3,000. Quite an expensive mistake for the tank commander.

We learned the next day that the tank was somehow responsible for accidentally starting the grass fire. The tank commander, wanting to be a good guy and fix what he had done, ordered his driver to drive over the burning grass to put it out. Bad idea. But it turned out to be good entertainment for us grunts.

Figure 22-1: The fire off in the distance is a tank burning. It caught fire trying to put out a grass fire by driving over it. As it burned, the munitions on board exploded one shell at a time. It was good entertainment for several hours.

Burning Tank

23. Food and Water from the Sky

Most days, we got our water from the small streams we crossed as we moved through the jungles on our search and destroy mission. As we crossed them, we usually took the opportunity to fill any of our empty canteens with water. Most guys carried three or four canteens. Once filled, we put one or two iodine tablets into each canteen, shook it, and put it back in the canteen holder on our belts. We were told to wait at least a half hour before drinking stream water, so the iodine tablets had time to kill any harmful microorganisms in the water. Lots of us carried small packets of Kool-Aid that we got from home. We usually poured that into our canteens to kill the bad taste of the iodine tablets and/or the water.

This day, we were in an area that did not have any streams at all. That meant all of us were either out of water or running very low by late afternoon. Whenever the company reported we were low on water like this, base camp would send us water in five-gallon cans as part of our re-supply that day (See Figure 23-1). That night we really needed water. We were also almost out of C-rations as well, so we needed them too.

Unfortunately, we had stopped our movement much later

in the day than we usually did. It was getting late in the afternoon, and we were still working hard building our nighttime perimeter.

A couple of helicopters were already in the air above us waiting for clearance to land, but the LZ wasn't done yet. Soon, it became evident that we would not be able to finish the LZ before dark, so we would not be able to land those choppers to bring us our supplies. Because our need was great, someone made the decision to have our supplies dropped into the perimeter from the helicopters hovering above the trees.

Instead of dropping them into the middle of the partially completed LZ, where we could clear out all soldiers, the two choppers hovered side by side and started pitching out cases of C-rations and full five-gallon cans of water right over the guys working on the perimeter.

We never got a warning about what was coming. Suddenly, large C-ration cases and five-gallon water cans were falling from the sky. "WTF" was the expression heard most coming from the troops on the ground as we scrambled to the nearest big tree. Hugging the trunk of the tree was, we thought, the best way to avoid being crushed by this assault from the sky.

I was lucky enough not to have a chopper directly above me, but I hugged a tree anyway. A few guys started yelling "Shoot them down!" or other such exclamations that were full of profanity toward the helicopters.

While it was happening, I had one thought about the ridiculous idea that a mother back home might be getting a telegram that said, "Your son died from a falling water can." Do they give Purple Hearts for that?

Finally, the crews of the helicopters ran out of stuff to throw out and the two choppers flew away. Left in their wake, scattered all across the perimeter, were many busted open C-ration boxes and mostly split open water cans which were now empty of water. Luckily, I don't think anyone was injured.

After gathering up the C's and a couple unbroken water cans, the NCO's had to work out a rationing system for the surviving water. Each of us got a little water, but not nearly enough to get us through that night and the next day.

The next day, despite trying to ration our water, everyone was out by midday. I spent all afternoon thinking almost exclusively about water. By the time the helicopters came with water into our completed LZ late the next afternoon, we were all very dehydrated. It was the thirstiest I had ever been or ever would be in my life. The NCOs made sure everyone got one canteen of water to drink before the rest of the canteens were filled.

I remember that before we got the water in, a couple of guys came up to me and asked if I had any water left. They really looked desperate. Unfortunately, I didn't have any myself and couldn't help them.

I don't know about the rest of the guys, but from then on, I never missed an opportunity to top off my canteens with water whenever I got the chance. I didn't ever want to run out of water again.

After that experience, I could now relate to all those cowboy movies I had seen in my youth where the cowboy was lost in the desert and dying of thirst. I knew exactly how that cowboy felt. It wasn't good.

Figure 23-1: When we were in an area without streams to use for filling our canteens, base camp sent out water in five-gallon cans. One guy from each squad would gather all the canteens from his squad members and head over to fill them from the water cans. Once you filled the canteens, you gathered them up and took them back to the squad.

Much Needed Water

24. Swamp Crossing

It was past midafternoon when our search and destroy mission brought us to a small, swampy lake in a large valley. Lakes were rare in the Central Highlands. This lake had a large grassy clearing next to it. The grassy clearing was large enough to easily hold a night perimeter for the company.

The fact that the area was open grass meant we wouldn't have to cut an LZ—it was a natural LZ. Because it was later in the day than we usually stopped, not having to cut an LZ would be a very good time saver. Word went out that this would be our night position.

We organized ourselves by platoon and were assigned the spots where each rifle team should dig its foxhole. Just then I was told I would be joining a couple of other guys from different squads to be the observation post (OP) on the other side of the small lake. An OP's job was to keep watch and ensure that the bad guys didn't sneak up on a busy and distracted company.

Just before we (the other OP guys and I) started across the lake, the guys that had started digging the foxholes made a disturbing discovery. The grass and dirt was only about six inches thick and it was laying directly on bedrock, not on the usual, easily dug, red dirt we found everywhere else in the Central Highlands.

As we OP's were leaving, we heard the NCOs and officers discussing how to handle this unforeseen development. It was now definitely too late to move to a new spot for a night perimeter, but we needed foxholes in case we were attacked. I wondered how this dilemma would turn out.

As we approached the lake to cross it, we could see from the weeds sticking out above the water, that it was very shallow all the way across its 50-to-75-meter diameter. It was more of a swamp than a lake.

We entered the water and began to walk across single file. I was the last of the three-man column. The water was about knee deep all the way across.

When we were about a third of the way across, I thought I saw something on the pants leg of the guy in front of me when it came out of the water as he walked. As he took another step and his other leg came up a bit out of the water, I saw something was definitely on his pants. Leeches. Big ones. At least three. When his other leg raised out of the water again, there were four more on that leg.

My first thought was, "Poor bastard, he's got leeches all over his legs." My second thought was, "Wait a minute, I'm in the same water." Sure enough, I looked down and saw my pants legs were covered with leeches as well. I think we all realized what was going on at the same time because our pace quickened considerably the rest of the way across the lake.

The leeches were a dark purple color and about three to four inches long. Very ugly. I was used to seeing the small, black land leeches that inhabited the jungle floor, but these water leeches were new to me. I couldn't believe there were so many of them in that little lake.

When we exited the water, we stomped our feet and all the leeches fell off our pants legs. We all checked and none of us had holes in our pants and, luckily, all of us had our pants tucked into our boots. No leeches got on our skin. We moved about 50 meters into the thick jungle on the far side of the lake and set up our OP.

Just before dark, the LP guys came across the lake to relieve us and we walked back through the water again. On our trip back, the leeches were still there, waiting for us, waiting for their ride on our pants. It was only slightly less creepy crossing that lake for the second time.

When we got back to the perimeter, I was impressed with the solution to the foxhole problem. Because bedrock was just below the six inches of dirt and grass, we couldn't dig down. So, the guys simply cut 18-inch squares of sod and piled them up to form above-the-ground foxholes. The thick sod walls for each foxhole would be more than enough to stop any small arms fire if we were attacked during the night. If sod houses were good enough for the pioneers, sod foxholes were good enough for us.

My team's foxhole was just at the edge of the lake. As I sat on guard duty that night, I tried to imagine how terrible it would be for any enemy soldiers crossing that lake to attack us. They would have to stay mostly submerged in the water to avoid our gunfire. If they did, the leeches would have a field day with them. I almost wished they would try to attack us.

The next morning, we pushed over our above-the-ground foxholes and moved on. I never had another experience with water leeches like that the rest of my year in Vietnam.

25. C-Ration Meals

A case of C-Rations contained twelve different meal combinations. Each meal box contained one accessory packet as well. Following is a description of the C-Ration meals that composed 90% of the meals I had in Vietnam (See Figure 25-1).

B-1 Units	B-2 Units	B-3 Units
Meat Choices (in sml can):	Meat Choices (in lg. can):	Meat Choices (in sml can):
Beef Steak	Beans & Wieners	Boned Chicken
Ham & Eggs, Chopped	Spaghetti & Meat-balls	Chicken & Noodles
Ham Slices	Beefsteak, Pot. & Grav.	Meat Loaf
Turkey Loaf	Ham & Lima Beans	Spiced Beef
Fruit:	Meatballs & Beans	Bread, White
Applesauce	Crackers (4)	Cookies (4)
Fruit Cocktail	Cheese Spread:	Cocoa Bev. Powder
Peaches	Caraway	Jam:

Pears	Pimento	Apple
Crackers (7)	Cake:	Berry
Peanut Butter	Fruit Cake	Grape
Candy Disc, Chocolate:	Pecan Roll	Mixed Fruit
Solid Chocolate	Pound Cake	Strawberry
Cream		
Coconut		
	Accessory Pack Contents	
Salt	Instant Coffee	Cigarettes (4)
Sugar	Chewing Gum (2)	Toilet Paper
Cream Substitute	Plastic Spoon (not inside)	Book of Matches
Four P-38 can openers came in each case of C-Rations	Heat tablets (or C-4) for cooking obtained separately	'Stoves' for cooking made from empty C-Ration cans

Figure 25-1: Typical 'supper' meal of C-Rations. This looks like I'm eating a B-3 unit because there are my beloved cookies and a can of jam on the sandbag to my left. The cooking 'stove' is on the far right. The very small box near the C-ration box is a packet of four cigarettes that came in every meal. I always gave my cigarettes to my buddies who smoked.

C-Ration Meal

26. Breakfast

Most days while out in the boonies, we usually got up early enough to have time to make ourselves a C-Ration breakfast before saddling up and moving out. After a month or so with the company, I settled on a breakfast that I made myself almost every day the rest of my tour in Vietnam — when there was time, that is.

C-Rations don't have a 'breakfast' meal option unless you count the 'Ham and Eggs, Chopped' meal as a breakfast — which I didn't. I collected the ingredients for my favorite breakfast from several different meal units. The ingredients I used were: one can of bread (B-3 unit), one can of jam (B-3), one can of peanut butter (B-1), one envelope of hot chocolate (B-3) and maybe a couple of crackers (B-1 or B-2).

I started my breakfast by first making one or two cooking 'stoves'. Stoves were made from an empty, small C-Ration can, with holes punched into it with a can opener (Like the kind you used to open beer cans before there were pop tops). The triangular holes (usually four) were punched around the sides of the bottom of the can for air. Three or four more holes were punched on the sides of the top of the can. The metal triangles from the top holes protruded slightly into the open top of the can. They served as rests for any can or food you put on top of the stove and prevented it from falling in.

Then I'd place a heat tablet on the bottom of the stove and light it. Heat tablets were part of the supplies regularly sent out to us. They came in foil packets and were like solidified Sterno. When they handed out C's, you usually could get heat tablets as well.

If heat tablets were unavailable for some reason, you'd go over to the engineers and ask them for a chunk of C-4 explosive to cook with. They typically had plenty of C-4 and were usually happy to give you some. C-4 didn't explode if you put a match to it, it burned. The first time I saw someone start to light C-4, I almost freaked out. The guy just chuckled and said don't worry, this doesn't make it explode.

When cooking with C-4 though, you had to be very careful. C-4 burned much hotter than the heat tablets, so it was very easy to burn your food with it.

Once the first stove was started, I'd put a large size C-Ration can almost filled with water on it to begin heating. Next, I'd open the can of bread with my P-38 can opener and, using the trusty hunting knife I always carried, cut the bread in half.

Then I'd start a second stove going, sometimes quickly made from the recently emptied bread can. It was time to make my toast. I'd take one half of the bread I'd cut, balance it on top of the second stove, cut side down, and start toasting it over a burning heat tablet placed on the bottom. While it was toasting, I'd open a can of jam and a can of peanut butter and mix them together thoroughly, using a C-Ration spoon. P, B, & J in the wild.

Soon it was time to take off the first half of bread (now

toasted) and put the other half on the stove for toasting. By then it was also time to make my hot chocolate because the water was hot. I'd open the packet of hot chocolate and pour it into the can of near boiling water. I'd mix it thoroughly with the spoon I had used to mix the P, B, & J—after first licking it clean of course.

Then it was time for the second half of bread to come off the toaster. Now I was ready to spread the P, B, & J on both halves of the toast and begin to eat. As I ate the toast, I'd wash it down with some tasty hot chocolate. There was usually some leftover P, B, & J, which I'd spread on a few crackers as a second course. Occasionally, if I was still a little hungry, I'd open a can of one of the cakes (fruit, pecan, or pound), which came from a B-2 unit, and have that for dessert.

I think that the whole process of cooking and eating this breakfast only took about 15 minutes. We found it was a rule that the more you messed up C-Rations, the better they were.

27. Punji Sticks

We were working in the Central Highlands on a normal search and destroy mission. The area we were working in was primarily wild jungle, but there were some villages nearby, with networks of trails connecting them to each other and to other areas.

It was a typical hot and humid day moving through the jungle. We were on a long climb up a shallow incline toward the top of the hill. We had been going up for an hour or so. When I got to within 100 meters or so of the top, the word was passed around to be careful of punji sticks.

Punji sticks are very, very thin cuts of bamboo, maybe 18 to 24 inches long, sharpened to a needle point. The VC and NVA sometimes booby trapped an area with them to injure or slow the progress of opposing forces. They were often stuck into the ground at an angle. Their purpose was to stick into the leg of an unsuspecting soldier walking through the area. Other times, they tied them to trees at various heights so they could penetrate your body or face if you walked into them.

As soon as I heard punji's were in the area, I began to move much more cautiously, paying special attention to the ground ahead of me. I did see a couple of punji's stuck in the ground pointing at an angle down the hill toward me. They were very difficult to see. If I hadn't been alerted to the danger, I probably wouldn't have noticed them at all. As I walked by, I stepped on them, pushing them flat to the earth.

Before I got to the top of the hill, the guy in front of me pointed to a branch of a tree about head high and then continued on. When I got to that branch, it took me a moment to see the punji stick pointed directly at my face. I wondered why the previous guy had left it there. I knocked it down as I passed.

By the time I reached the top of the hill, soldiers were already busy clearing out an area so helicopters could lift out a couple of our guys who were wounded by the punji sticks. As I walked into the work area, I saw the two wounded guys sitting against trees waiting to be evacuated. Both had their pant legs cut open and pulled up over their knees. Both had wounds in their lower leg and their injured calves were badly swollen.

The two wounded guys were from first platoon, so I didn't know them.

Thirty-seven years later, I met a new guy at one of our battalion reunions. When we met, I told him I didn't remember him at all, even though we were both in Charlie Company at the same time. That was not unusual. In Vietnam, there was very little time to socialize with other members of the company who were not in your squad or platoon.

During the reunion, the new guy and I were in my car making a beer run. We were, naturally, talking about Vietnam. I asked him if he was ever wounded while he was there. He said that he and another guy were wounded by punji sticks once.

I nodded and smiled. Then I told him I remembered seeing

him that day as he waited to get evacuated out. He and I, and our wives, became very good friends from that reunion forward and shared many good times together.

28. The New CO

In January of 1967 I felt I was finally getting well-adjusted to my life as a rifleman in the jungles of the Central Highlands in Vietnam.

One day as my squad was coming back into a patrol base after a patrol, we were met at the perimeter by the Company Commander (CO) and a couple of NCOs. As we filed past them, the CO stopped me and asked if I was PFC Witt. Surprised, I answered, "Yes, sir." He then asked me what I knew about radios. I told him I had had 2 ½ hours of training on radios in AIT. He simply said, "Good." I sensed that was all he wanted, so I continued into the perimeter and then to my bunker.

Once there, I found my platoon sergeant, told him what had just happened, and asked him if he knew what that was all about. He told me that Charlie Company was getting a new Commanding Officer in a few days. The old CO was taking his two RTO's with him to his new assignment at base camp, so the new CO needed two new RTO's. The sergeant said I was going to be one of the new RTO's.

I told the sergeant I didn't want to be an RTO. He smiled

and told me I was picked for the job, and I had no choice in the matter.

As the day for the Change of Command ceremony approached, I heard stories about the incoming CO. He was the current CO of HHC back at base camp and had a reputation of being a real bad ass. He made life miserable for any screw up there and especially for any infantryman who tried to find excuses to stay in base camp rather than be with his infantry company in the field. His methods were so successful, many guys found, because of him, they preferred to be out in the boonies humping the hills over being back in the cushy and significantly safer life at base camp.

With his reputation, everyone was full of dread anticipating the arrival of the new CO.

The change of command ceremony was held at a large fire base. Charlie Company was given a few stand down days to rest while we were there. This meant we had no perimeter guard or patrol responsibilities. We had large tents that served as barracks to sleep in and as the company headquarters.

The change of command ceremony brought us a new CO (See Figure 28-1). At the same time, by default, I officially became one of the new CO's RTO's. Oddly, that fact which impacted me so much, was not even mentioned during the ceremony.

After the ceremony, the other new RTO and I were not introduced to the new CO. But the First Sergeant did take us aside and told us to move our gear into the command group area. He also told us that, despite Charlie Company's stand down, as RTOs we still had to monitor the radios in the company headquarters tent.

We worked out a schedule for radio watch and I was assigned the late evening shift.

It was about 21:00 that evening and I was sitting at a table writing a letter as I monitored the radios. I was the only person in the large tent at the time. Suddenly, in walked the new CO, this bad ass, hell raising man that everyone feared. I immediately stood up and was very tense, not knowing what to expect.

He looked around, then asked me, "What's your name?" I told him. "What are you doing here?" "Radio watch, sir." Still checking out the tent, he asked, "Where are you from?" "Wisconsin, sir." He hesitated a moment then asked, "What part?" "The southeast part, sir." Another hesitation. "What town?" I responded, "Kenosha, sir."

With that, he lowered his head a bit and began to slowly shake it back and forth. My immediate thoughts were, "What the hell did I do? What did I say? Am I in trouble? What?"

Then he quietly said, mostly to himself, but out loud, "I don't believe this. The first enlisted man I talk to in my new company is from the same hometown as my wife and where she is right now."

When I heard him say that, my mind started going a thousand miles per hour wondering what it could mean for me. I quickly decided—this might be good.

We chatted a while about Kenosha where he had spent a lot of time and found there were a lot of things we were both familiar with. It was a very nice conversation and good start to our time together. Later on, our families in Kenosha even got together a few times while we were still in Vietnam.

As time went on, the whole company and I found that he was an excellent officer and a great company commander. He was not at all like the officer he was in HHC. His persona as

HHC CO was just an act to get the most out of the soldiers in base camp. During the six months we served together, we became friends, at least as well as a Captain and a SP4 can be friends (See Figure 28-2).

The new CO's name was Captain Ed Northrop.

In 2003, we renewed our friendship at one of the reunions our battalion holds and have since become very close. Now, we and our wives get together at least once a year somewhere in the country to spend a few days with each other. Lots of 'Vietnam stories' are always told during these get togethers. Most of them are mostly true. Lots of laughs as well. I will still call him Captain sometimes, even though he retired a Lt. Colonel, and he occasionally likes to call me "Private!"

Ed retired from the Army after 20 years of service. After his retirement he went on to become a very successful entrepreneur, eventually owning several Burger King franchises as well as other enterprises.

In recent years, he has served as the honorary colonel of the 12th Infantry Regiment. He was also the leading force in creating the 12th Infantry Regiment Monument on the 'Walk of Honor' on the grounds of the National Infantry Museum at Fort Moore (formerly Fort Benning), GA. I was very honored to contribute in a small way to the success of that project as well.

His friendship and the friendships I've shared with the other men who served with the 1/12th in Vietnam are the best things to come out of the Vietnam War for both my wife and me.

Figure 28-1: Picture of Charlie Company's CO taken before he became our CO. A great officer and great friend. This picture was taken just after the 4ID arrived in Vietnam. On the ship (the USS Walker) bringing them to Vietnam, the soldiers got the idea to all get Mohawk haircuts before arrival. Our CO joined the men. I was not part of the original group of 4ID soldiers on that ship. I joined them as one of the first replacements about four months after they arrived in Vietnam.

New CO

Figure 28-2: The CO is in the middle, and I am on the right. The soldier on the left is the Charlie Company executive officer at that time. He was also my first platoon leader and, later in life, a good friend to both the CO and me for many years. We are waiting to do an air assault. Other men can be seen staged in the background waiting for a mass of helicopters to come in and pick us all up at the same time.

The CO and Me with Exec Officer

29. First Stream Crossing

It was January 1967. I was the new RTO for the new commanding officer (CO).

The day before, we had a change of command ceremony where this new officer took over as our company commander. When the previous CO left, he had his two RTOs transferred to base camp to continue working for him there. Another soldier and I were selected to be the new RTOs for the new CO. I had never expected to become an RTO and was more than a bit apprehensive about my new duties.

Charlie Company was just leaving the brigade firebase to start our first search and destroy mission under the new CO. As soon as we entered the jungle, the new CO stopped and motioned me and the other RTO to come over to him. When we did, he looked us both in the eyes and very seriously said, "The most important job each of you has is to stay close to me. If I need a radio, I'd better have it in my hand immediately, understand?" Both of us nodded meekly. The new CO had the reputation of being a real hell on earth to the troops he previously commanded back at base camp. Neither of us wanted to get on his bad side.

Then, he surprised us by saying, "The second most important job each of you has over the next few days is to tell me when I'm f..king up. I'm new to this job and I don't know exactly how things have been done out here in the past. If I give an order that either of you thinks is really bad, you have to tell me, OK?" We both responded with, "Yes, sir."

I walked away from him, not too far of course, thinking, "That's a good man. He understands he doesn't know everything."

Later that morning as we were traveling through the jungle, I was walking in front of the CO and the other RTO was walking behind him. We came to our first stream crossing. The stream was maybe 30 meters across, but only deep enough to cover our boots, and was full of rocks jutting out of the water. As with every stream I had crossed in the last couple of months, I just walked through the water to the other side. I was accustomed to wet boots. Out here, wet boots were unavoidable.

When I got to the other side, I stopped and looked back to make sure the CO was right behind me. You know, not too far away as ordered. Instead of seeing him halfway across the stream as I expected him to be, I saw him standing on the far edge of the stream intently looking down at it. What the hell was he looking at? Then he hopped to the top of the first rock sticking out from the water near the far shore. He stood on that rock and continued studying the rocks ahead of him. Oh my God, I realized this hell raising, rock hard, airborne ranger was trying to cross the stream without getting his booties wet!

We always traveled through the jungle very spread out, so you generally couldn't see more than seven or eight guys at any one time through the thick undergrowth. However, at stream crossings, you would usually see several more guys than that.

146

I glanced up and down stream and there were probably 20 or more men in view. All of them were stopped, watching the new CO, apparently trying to cross his first stream without getting his boots wet.

The CO hopped to the second rock, then the third. He was now successfully about a quarter of the way across, and his boots were still dry. On his fourth jump, he ran into trouble. I think he thought he had to make a quick two-rock jump because the rocks were a little further apart. He didn't make it.

After that fourth jump, his foot slipped, and his momentum started to carry him forward. He was definitely not going to make it. After two or three splashy steps, he went down, face first, spread eagle in the shallow water. I think everyone within sight turned away to hide their smiles. I know I did.

When I looked back again, the CO was picking himself up out of the water. The whole front half of his body was wet. When he regained his feet, he kept his dignity and simply brushed off some of the water from his front and walked on toward me—splashing through the water with his now wet boots. I don't think he looked around at any of the witnesses scattered around him. I thought the whole event was hilarious. So much so, I turned and walked very quickly into the jungle so he couldn't see me smiling from ear to ear.

I suddenly realized I'd better slow down or I'd be break-ing order number one. I stopped. When he caught up, nothing was said. After all, 'nothing' had happened, right? We simply continued on through the jungle—Charlie Company and their new CO, who now not only had wet boots, but wet pants and shirt as well.

The CO turned out to be a great CO. I learned a lot from him and about him. I learned the welfare of his men was always the most important thing to him. I learned he was always thinking. I learned he thought best when he was doing something with his hands. He always kept busy doing something while the rest of us were building our night perimeter. I also learned that every stream we came to after that first day, he crossed by simply walking through the water like the rest of us.

We became friends over the next six months. When it came time for him to go back to the States, he arranged for me to get transferred to the battalion firebase to become the battalion commander's RTO. That provided me a significantly easier and somewhat safer life for the last three and a half months of my tour in Vietnam.

I am proud to still call our CO my friend.

30. 'Friendly' Fire

That day, battalion told us that there were NVA elements just ahead of us as we moved through the jungle. Our job was to close with them and engage. Since battalion was so convinced we were just about to make contact, they decided to bring in air and artillery support prior to our making contact with the enemy.

As we moved through the jungle, battalion had us throw smoke to identify our position to the FAC (Forward Air Controller) flying overhead. They then told us that aircraft would be strafing the jungle just ahead of us to soften up the NVA.

When the strafing runs began, I was surprised at how close they came to us. Our point element quickly radioed back to the CO that they were much too close. The CO relayed that to the FAC, and the strafing runs were adjusted further to the front.

After several strafing runs were made, battalion told us that they were switching to artillery support. Word went out to expect several volleys to our front.

Shortly, several volleys of artillery rounds landed, not to our front, but right in the middle of Charlie Company.

The rounds screamed a horrible sound as they came in. It is

like a high pitch, extremely loud, Eeeeeee. Then the explosion. Since we were in the thick jungle, most of the rounds exploded in the treetops, sending shrapnel everywhere.

When the first rounds came in, all I could do was dive to the ground like everyone else. There was no cover anywhere. As I lay on the ground, I kept trying to get further up inside my steel pot. But as hard as I pushed, the only thing that I could get in was my head. The rest of my body just lay there, exposed on the jungle floor.

After three volleys, the CO finally convinced them that we were taking friendly fire and they should cease firing. I think they argued with him momentarily that we must be getting hit with rockets from the enemy, not artillery. However, we all knew it was not rockets that came in on us that day.

With that, our advance was called off. We had wounded to attend to. I think we had three or four guys hit by shrapnel. We were very lucky because it could have been so much worse.

I gained three things out of this experience. The first was a deep appreciation that we did not have to face enemy artillery. The NVA just didn't have the ability to bring artillery into the Central Highlands to use against us. Mortars, yes, artillery, no. Artillery coming in on you was terrible.

I also got a much better understanding of why guys in past wars got shell shocked by constant bombardment by artillery. We only had that one experience, and that was bad enough. I hated to imagine what day after day of artillery coming in on you would be like.

The third thing I gained, was a deeper respect for our NVA enemies. They had to face what we just faced, all the time, and it would be many, many times worse. I realized they were very good soldiers to keep fighting so tenaciously against us as they did.

31. What I Carried

Pack Items	How Many	Weight Each	Total Weight
Rucksack Frame	1	3	3
PRC-25 Radio	1	22	22
Long Antenna and Case	1	2	2
Extra Battery	1	3	3
M-16 Magazines (18 Rounds in Each)	21	0.7	14.7
Claymore Mine bag for Carrying Magazines	1	1	1
M-26 Fragmentation Grenades	8	1	8
Smoke Grenades	4	1.5	6
Machete	1	3	3
Entrenching Tool	1	5	5
Poncho & Poncho Liner	1	3	3
Rubber Air Mattress	1	3	3
Mosquito Net	1	1	1
Canteens	3	2	6

Pistol Belt w/ Ammo Pouches & First Aid Packet	1	4	4
Small Canvas Pack w/ personal gear (toiletries,	1	5	5
Stationary, washcloth, ½ towel, extra socks,			
A few pictures, some letters, a camera, and a			
Miniature bible)			
Total for Pack			89.7
Other Gear			
M-16	1	7.5	7.5
C-Rations (1 days' worth)	3	1.75	5.25
Soda's	2	.75	1.5
Helmet	1	5	5
Boots	1	2	2
Clothes	1	3	3
Maps for CO	1	.5	.5
Hunting Knife	1	.5	.5
Total Other Gear			25.25
Grand total of all gear			114.95

See Figures 31-1 and 31-2 Below

Figure 31-1: My 89 pound pack and 25 pounds of other gear. Just behind my right shoulder is the speaker for the radio. We could use the speaker or the handset for listening to incoming transmissions. We usually just used the handset to keep the noise level down. Under the poncho roll is a claymore mine bag filled with most of my 21 magazines of ammo. Below that is a small canvas pack for my personal gear. I got rid of the original nylon rucksack bag because it was too big. I found, if I had room in my pack, I'd fill it with something. It got way too heavy. So, over Christmas rest at base camp, I found this smaller pack which somewhat forced me to limit what I carried. Also shown are two of my canteens. My fatigue pockets are filled with lots of things. It looks like my pant leg pocket is filled with maps for the CO. The antenna for the radio is visible, sticking up in the air.

What I Carried (Side View)

Figure 31-2: One more canteen can be seen on the left. Also, lots of stuff is in the other pant leg pockets. Probably C-Ration meals. An entrenching tool handle is hanging down my left side. Hidden somewhere on the pack is my trusty machete. On the front, which isn't shown, I carried two ammo pouches attached to my webbing that were filled with fragmentation grenades (8) and had smoke grenades (4) attached to them.

What I carried (Back View)

32. Screaming in the Ia Drang

In the spring of 1967, our area of operation was in the Ia Drang Valley of Vietnam.

The area we were in was thinly treed grasslands and fairly flat. The brown grass was approximately two to three feet high. We were moving through the area on a search and destroy mission, looking for the NVA.

There were approximately 100 men in our company. As was our standard operating procedure, the company was moving in seven columns shaped somewhat like a diamond. We always kept good separation between soldiers (usually 10 to 15 meters) within our formation to avoid high casualty counts in case of an ambush. As a result, our formation was very spread out as we moved through the grasslands.

As usual, the RTO that carried the company radio was walking in front of the CO and I (carrying the battalion radio) was behind the CO. We were walking single file roughly in the middle of the center column of the company formation.

All was normal when, suddenly, we heard men screaming from far in front of us. We all stopped and looked at one another. The screams were a mix of yelling and screams of pain. There were no weapons firing. What was happening up there?

My first thought was that our men must be involved in hand-to-hand combat. Poor bastards! There was a moment of indecision on everyone's part—charge to the front to help or take up defensive positions? Just then, one loud, clear scream told everyone exactly what was happening to our lead elements, "Beeeeees!"

At that moment, everyone's first instinct was "Retreeeeat!" However, we bravely held firm where we were and waited.

The screaming came closer, but the men under attack finally managed to outrun the enraged bees. Once the men were out of the immediate area of the beehive, the bees stopped attacking them. Thankfully, the bees never got to our section of the company formation.

Once the bee attack ended, we had to deal with the aftermath. Several of the men while they were under attack and fleeing for their lives, ripped off their heavy backpacks and abandoned them. Some even dropped their weapons. The NCOs were not happy with those soldiers. Everything stopped until the men who were missing their equipment went back to the front to retrieve it. Slow and easy seemed to work, as the bees didn't renew their attack on them.

Once everyone was whole again, the CO decided that a detour around the bees was the wise choice. He didn't want to chance a renewed attack from the bees if we went the same way.

We didn't suffer any permanent casualties that day. Apparently, the men who were stung were not overly allergic to bee stings.

One time, after a different bee attack, one of the soldiers who was stung was very allergic. He started to go into anaphylactic shock. We had to call in a chopper to lift him out of the jungle by sling to get him back to base camp for treatment. He could have died, but thankfully didn't.

There were lots of ways to die in Vietnam.

33. 501 North — 15-16 February 1967

Our day started out, as most did in Vietnam, by waking before dawn.

The day before, we had just moved back into one of our former fire bases, the fire base known as 501 North. Luckily, we had enough time that day to rebuild the perimeter and command bunkers before nightfall came. When we left it three months before, we just collapsed in all the bunkers and left all the sandbags filled. All we had to do that afternoon was simply rebuild the bunkers using the already filled sandbags. A real time saver.

However, there was not enough time to also cut down the thick vegetation that now occupied our previously clear fields of fire around our perimeter. Undergrowth, several feet high, had grown in the three months since we last occupied the fire base, and came to within 10-15 meters of the front of the bunkers.

Oddly, this fire base was the very first fire base I had ever seen in Vietnam, but that was three months earlier. A couple of weeks after I joined Charlie Company in the jungle the previous November, our turn came to pull 'palace guard'. 501 North was the fire base we went to and guarded. I remember walking out of the jungle and seeing a fire base for the first time and thinking to myself, "This isn't exactly the 'palace' I had pictured in my mind." Little did I know then what the future had in store for us and for this place.

Shortly after rising that morning, I decided to make my breakfast as daylight began to break. Looking for a good place to sit while making my traditional breakfast of hot chocolate and peanut butter and jelly C-Ration toast, I casually walked from the command bunker over to a large hole located on the perimeter. It was just past one of 2nd platoon's bunkers. I sat down with my feet dangling in the hole and started heating water for my hot chocolate. While it heated, I opened my can of bread, cut it in half, started toasting it, and then opened the cans of peanut butter and jelly and mixed them together. When the water was hot enough, I mixed in the chocolate, tasted it, and remarked to myself that this was the best damn hot chocolate I had ever made. I had just put it down when the morning silence of the jungle was broken.

Suddenly, the sound of heavy small arms fire was coming from outside the perimeter, and down about five or six bunkers from where I was sitting. I immediately jumped up and began running for the command bunker, which was probably about thirty meters away. But, as soon as I started to run, very heavy, incoming small arms fire began from the jungle directly in front of where I had been sitting. I could hear the bullets going by and knew I'd probably never make it to the command bunker, so I dived for cover behind the 2nd platoon bunker on the perimeter.

There were four guys already in the bunker, armed with an M-60 machine gun and three M-16's — all were friends of mine from second platoon. Firing was very intense in both directions for several minutes. I yelled to the guys inside to let them know

that I was just behind the bunker. They asked for some ammunition bandoliers that were laying near me, and I passed them
in to them.

There I was, a major battle starting, my first, I'm not at my
post as the CO's RTO, and I'm unarmed—my M-16 was about
25 meters away, leaning against the side of the command bunker. I'm also not in a bunker, no room for a fifth guy in this one,
I'm behind it and wondering what's coming next. It wouldn't
take long to find out.

The area directly behind the bunker was the sleeping area. It
was surrounded by three heavy logs lying on the ground. Less
than five minutes after the battle began, I realized how close in
the enemy really was. An NVA grenade came sailing through
the air, over the bunker in front of me, passed by my face, and
landed just on the other side of one of the logs next to me. I
yelled "Grenade!" and laid down as flat as I could behind that
log. The grenade exploded less than two feet from me, but the
log did its job and protected me from harm.

The guys quickly started passing their grenades out to me.
I could throw from out there and they couldn't from inside the
bunker. I immediately threw three grenades over the top of the
bunker toward the direction I thought the NVA grenade had
come from. But, because I was behind the bunker and really
couldn't see, the direction and distance were just guesses.

A couple of minutes later, I heard a muffled explosion and
then screams from inside the bunker. It took a moment for me
to realize what had happened. An NVA grenade had managed
to get into the bunker and had exploded at the feet of the four
soldiers inside.

All the guys were wounded, with shrapnel in their legs. All
came crawling out of the side entrance of the bunker to where

I was in the back. I yelled for a medic and then threw five grenades in a spray pattern in front of the bunker. Someone yelled that there was no one in the bunker, so in I went. Inside the bunker was an M-60 machine gun and three M-16's. I immediately tried to use the M-60, but it was jammed—it didn't work the rest of day—so I began firing using one of the M-16's (See Figure 33-1).

In the meantime, one of our medics, God bless him, God bless them all, ran under fire to our bunker and began helping the wounded. Three of the guys were evacuated away from the perimeter, but one of the guys, Daniel, felt that he was still capable of helping, so he came back into the bunker with me. I'm still very thankful he did. He was a very big help that morning. He couldn't stand very well with his leg wounds, so he sat or leaned against the side of the bunker. He loaded magazines for me for several hours.

At that time, the Army still didn't have a lot of magazines available, so most of our M-16 ammo was carried in bandoliers still packed in 20-round cardboard boxes. As the battle went on, the ammo had to be loaded into the limited number of magazines we had. Daniel would load and I'd shoot. About an hour into the battle, a medic brought me my M-16 when he came to check on Daniel again. I don't know why, but I felt more comfortable when I finally had my personal weapon with me in that bunker.

Most of the time I was just shooting at movement in the undergrowth, just quick shots in the direction of the movement, never really having time to aim. I never set my M-16 on full auto. I purposely didn't use full auto, because I felt I had to maintain a good volume of fire coming out of that bunker, but, at the same time had to conserve ammo. It was good I did.

Because, by the time I left that bunker late that afternoon, I wasn't counting how many magazines I had left any more, I was counting how many bullets. I had used up almost all the M-16 ammo carried by three soldiers.

One time, I clearly spotted an NVA soldier through a small crease in the undergrowth. He was crouched down, but I could see him from his belly to the top of his head. He seemed to be talking to someone to his left. I remember thinking "This guy is mine". I took slow, careful aim at his forehead and squeezed off a round. I couldn't believe it when he didn't fall! He should have been dead with a bullet in his head!

Before I could gather myself and shoot again, he moved to the side, and I couldn't see him anymore. I've always felt that I probably should have switched to full auto for that shot. (A few days later we had a mad minute and I discovered that my M-16's sights were way off. It was shooting about a foot over where I was aiming. Ever since, I've blamed the weapon and not my aim for the miss.)

In the end, it turns out that the only man I ever deliberately and methodically tried to kill, I missed. I'm OK with that now. But at the time I was very angry that I'd missed him. I'm pretty sure it wouldn't have bothered me if I had killed him — it was war — and he was trying to kill me. But, since I did miss him, I never had to deal with his death and therefore will never know for sure if it would have bothered me or not.

The next day, we found eight dead NVA, 20 to 30 meters in front of my bunker. But I don't know who killed them. Could have been me with the grenades or M-16, or it could have been the guys who were in the bunker before I went in. I'll never know. If it was me, fine. If it wasn't, fine as well. I've never had a problem with it.

Later in the morning, the shooting started to quiet down a bit in my area and our air cover was working hard. I noticed that every time an aircraft passed over a particular area, about 200 meters outside of the perimeter, a heavy automatic weapon fired on it. The direction to that automatic weapon was exactly the same direction from which we had moved in the previous day. I decided to tell the CO about it. I told Daniel I'd be right back and ran to the command bunker.

When I got there, the CO asked me where I'd been all morning. I think he thought I was dead or wounded. I told him I was in a bunker on the perimeter and had to go back, but just needed to tell him about the NVA automatic weapon. I told him the direction, guessed at the distance, and then ran back to my bunker. The CO relayed the information to a FAC (Forward Air Controller). A few minutes later, an aircraft dropped a very BIG bomb exactly where I thought the gun was at. It must have taken the machine gunner out because I didn't hear him shooting again the rest of the day.

As far as I could tell from the volume of shooting around me, most of the morning action was directly in front of and to the right and left of the bunker I was in. That's why I was so surprised by what happened later that afternoon. I saw two helicopters shot down that day. One was a Huey filled with troops, the other was the battalion commander's small observation chopper.

The Huey was one of a group of three attempting to come into the perimeter with reinforcements from, I believe, Bravo Company. The problem was, they chose to approach the fire base from the direction that would take them directly over what I figured was the strength of the enemy. When the helicopters dropped to within 75 to 100 feet of the ground and were just about over the perimeter, the whole jungle opened up on them. Two peeled off and flew away. But one just rocked back and forth, and then tipped sideways in the air and crashed inside the perimeter (See Figure 16-2).

I never saw it hit the ground because I knew it was going to crash pretty close behind my bunker and felt that the bottom of the bunker was the place to be when it hit the ground. I think one of the helicopter crew was killed and most of the rest were badly injured in the crash. I learned later that both of the other choppers also went down before making it back to their base.

The observation chopper, carrying the battalion commander, was kind of the same story. I guess the Colonel wanted to be on the ground during the battle, so he directed the pilot to land. Unfortunately, he chose to land just outside of the perimeter and in the same general area that the Hueys flew over.

I watched as the little chopper approached, not believing they were actually going to land there. When they were about 30-40 feet from the ground, predictably, the jungle opened up on them. I saw both men exit the chopper BEFORE it hit the ground and run for the nearest bunker on the perimeter. I heard later that the Colonel had to spend quite some time in that bunker before he could get out.

As the pilot jumped out of the helicopter, he didn't bother to turn it off. So, when the helicopter hit the ground, it continued to run and run and run and run. It must have run on its own for an hour or so. I think both sides were shooting at it in the end just to STOP the nerve wracking, Tha-thump, Tha-thump, Tha-thump, sound of its engine. Finally, someone was success-ful in killing that small helicopter because it quit running. I personally was glad that it was finally stopped.

It was probably about noon when one of the guys, named

Phillip, in the bunker next to mine, yelled over and asked how we were doing. I yelled back that Daniel was getting weaker. Since there were four guys in his bunker, Phillip decided to run over to join us, which he did. A short time later, some of the medics ran to our bunker and helped evacuate Daniel to the interior of the perimeter. Thank you, Daniel, for your help and your bravery. I couldn't have made it without you.

Fortunately, our area of the perimeter was a lot quieter in the afternoon than it was the morning. About the middle of the afternoon, I had to answer a call of nature — number two — but obviously still didn't want to leave the bunker to do so. The solution came in the form of an M-60 ammo can, which I used, closed tightly, and tossed out of the bunker. I now understood why all those toilets in basic training barracks don't have walls around them. Sometimes in combat you must 'go' with your buddies nearby — basic training just started getting you used to doing it.

Later in the afternoon, both Phillip and I became very hungry. My delicious hot chocolate was still sitting on the ground nearby, but I decided it wasn't really worth going out to try to get it. Then we remembered that battalion had sent us our mail the previous day instead of our mortars (which we could have really used during the battle). The mail included packages, and there was always food in packages. We checked behind the bunker and, sure enough, we found a package. However, the only food item left inside it was a Jiffy Pop Popcorn. So, as I watched for the enemy, Phillip popped the popcorn while sitting in the bottom of the bunker. That popcorn was the only food I had that day (See Figure 33-3).

One vivid memory I have from that morning is the sound of the voices immediately after the shooting started. The NVA were obviously not quite ready to attack yet when our morning sweeps surprised them. Just after I dived behind the bunker, I could hear the jungle filled with voices of what were probably the NVA officers and NCOs yelling directions to their men. I remember thinking that it sounded like there were a whole lot of them out there and they were very close. It was very haunting, strange, and surprising to me, to hear all their voices mixed in with the sounds of small arms fire. I guess in all my imaginings about what a battle would be like, I just never thought about, nor expected, to hear the voices of the enemy in combat.

Another memory I have concerns the wonderful air support we had that day from both Army gunships and Air Force planes. The FAC supporting us that day did an incredible job. The war birds dropped their rockets and bombs very close to our perimeter all day long, but it was absolutely necessary. They did a great job. Really put on a show (See Figure 33-4).

I also remember the sound of an M-40 grenade launcher being fired over and over by an NVA soldier. He must have captured it and was using it against some of the bunkers to my right. I'm not sure if he ever hit anything with it, but he sure kept trying.

Another memorable event happened later in the morning. The guys in the bunker to my left had spotted an NVA soldier, 20 to 30 meters outside our perimeter carrying an RPG. They shot him. They figured he was down, but the RPG was still out there, with the potential of being used against them by another NVA soldier. One of the guys, I don't know who he was, volunteered to go out to get the RPG. They yelled over to me to warn me that he was going out.

Soon, out he came, carrying only his M-16 and at a full run. He ran through the undergrowth, found the RPG, grabbed it, and came running back in. I took his picture as he was running back to his bunker with the RPG in his hand. I've never found out who he was and have never been able to give him a copy of that picture of his heroics (See Figure 33-5).

Just before dark, guys from Alpha Company came up to our bunker, and said they were there to relieve us. I took my M-16 and went back to the command bunker to resume my duties as RTO. I stayed up with the CO the rest of the night, although I honestly don't remember any of it. I finally got some sleep late the next morning (See Figure 33-6).

A few days after the battle, the after-action intelligence report stated that Charlie Company was attacked that morning by a reinforced battalion (500 to 600 soldiers) of NVA regulars. Their mission that day was to kill an entire American company. They chose Charlie Company. We only had about 100 soldiers at the start of the battle, but they still did not accomplish their mission.

The tenacious defense of 501N by Charlie Company allowed both Alpha and Bravo companies to be successfully airlifted into 501N to reinforce Charlie Company by midafternoon. The three companies, with air support by the Air Force, successfully drove off the attackers after inflicting heavy losses on them.

During this battle, our battalion lost 13 soldiers KIA (five from Charlie Company) and many wounded. After the battle, we found well over 100 dead NVA soldiers around the firebase. There was no way to know how many more of their dead and wounded were taken away by the NVA as they withdrew from the battlefield and fled into Cambodia. A couple of wounded NVA were captured and brought into the firebase for treatment the next day (See Figure 33-7). Eventually they were shipped out to a POW camp.

Finally, there is one thing about that day of which I am absolutely certain. I know that the morning sweeps we sent out actually surprised the enemy before they could surprise us. That fact undoubtedly saved my life. Louis Willett, who was KIA that morning and posthumously awarded the Medal of Honor, and his fellow soldiers on that sweep, gave me just enough warning to get to cover.

I am positive that as the NVA got ready to launch their surprise attack that morning, there were several NVA soldiers watching me from the edge of the jungle. I know that each had decided that as soon as he received the order to fire, he was going to take out the skinny blond guy cooking his breakfast. If not for Willett and the rest of the guys on the sweeps giving me those few seconds of warning, I know I would have died there next to my hot chocolate and been one of the first casualties of the battle for 501 North.

Thank you, guys. Thank you, Louis.

Figure 33-1: This is the view from my bunker during the battle of 501 North showing the tall weeds that were much too close to the bunker. Unfortunately, the field of fire was not completely cut the night before and the NVA were able to sneak up to within about 15 meters of the bunker without being seen. That allowed them to throw grenades that reached the perimeter bunker. Normally, the field of fire was cut to 25 to 30 meters away from the bunker. That morning, there were many NVA soldiers hidden in the weeds directly in front of the bunker.

View From My Bunker

Figure 33-2: This helicopter was shot down and crashed in the middle of our perimeter the afternoon of the battle. I believe one young crewman died in the crash. The rest of the crew and other soldiers onboard were all injured. Years later, I met and became friends with one of the soldiers who survived this crash.

Downed Huey

Figure 33-3: This Jiffy Pop popcorn was the only meal I had that day. It was late afternoon and the shooting had died down quite a bit in our section of the perimeter. We found the popcorn in a care package behind the bunker. My bunker mate popped it in the bottom of the bunker while I watched for the enemy.

Only Meal of the Day

Figure 33-4: One of the many, many bombing runs done by Army helicopters and Air Force planes throughout the day supporting, first just Charlie Company, and then the rest of the Battalion later in the day. Many of the bombs, rockets, and napalm were dropped very close to our perimeter because the enemy was in so close. Several times, small shrapnel shards hit the bunkers along the perimeter. A few times, we felt the intense heat from the napalm as well.

Gunship Support

Figure 33-5: This brave soldier ran out to recover this RPG during a lull in the battle. He and his bunker mates did not want the NVA to recover it and use it against their or someone else's bunker.

RPG Recovery During Battle

Figure 33-6: This is me after being up for about 30 hours. When I was relieved from my bunker on the perimeter by members of Alpha Company, I joined the CO in the command bunker and stayed up all night with him as he directed air and artillery attacks on the enemy. That morning, the enemy broke contact and disappeared into the jungle and probably back into Cambodia. They left behind many dead comrades scattered around 501 North.

Morning after Battle

Figure 33-7: The morning after the attack on 501N, our sweeps outside the perimeter found this WIA enemy soldier. He was treated where he was found and then he was brought into the firebase for additional treatment (seen here). He was eventually choppered back to a hospital and taken to a POW camp.

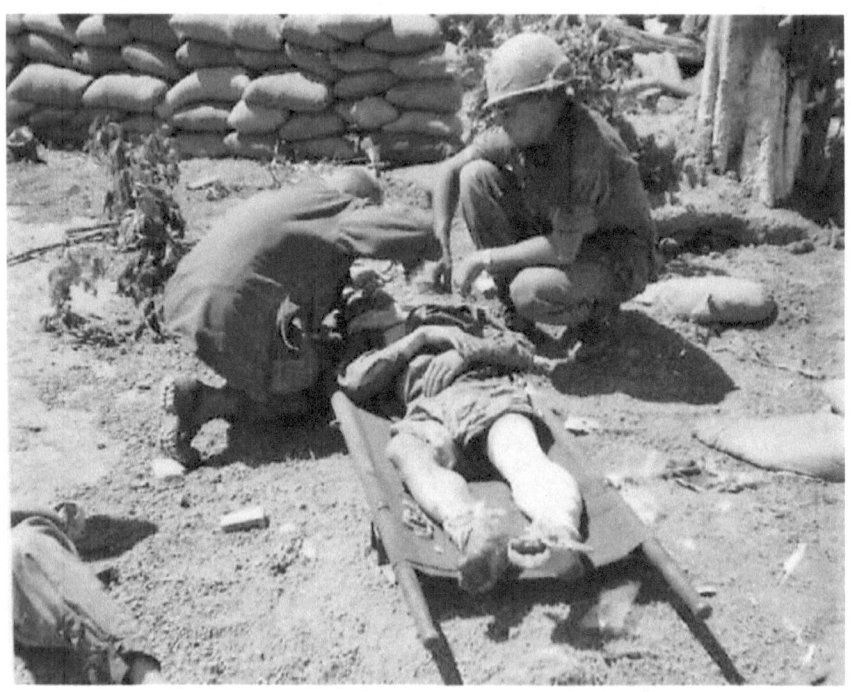

WIA Enemy Soldier

34. Five Snake Stories

SNAKE IN THE CROTCH:

We were busy building our night perimeter, which involved many tasks. Among them were digging foxholes, cutting overhead cover for the foxholes, cutting a landing zone, and clearing our fields of fire.

The triple canopy jungles of Vietnam contained many different types of vegetation and trees. One of the dominant types of trees was a species that had smooth, creamy white colored bark and extremely hard wood. The young ones grew straight and tall and usually didn't have any branches for the first 20 to 30 feet of the trunk. Those kinds of trees, with an eight-to-nine-inch diameter trunk, were ideal to use for overhead cover for our foxholes.

The mature version of those trees grew from 80 to 150 feet tall. At ground level, the whole circumference of the trunk was a series of large 'V' shaped crotches made by the roots which grew out several feet from the trunk. The 'V' shape of the roots could go up two to three feet above the ground.

This particular evening, I was standing next to one of those large, white trees using my machete to cut up a smaller tree for overhead cover.

As I was cutting, a friend of mine was walking by. He stopped maybe 10 feet from me. He said my name and then said, "Come over here." I stopped cutting and looked at him. "What?" He again said, rather firmly, "Come over here." Confused, I did what he said.

When I got to him, he pointed back at the place where I had been standing and said, "Look." I did and saw the object of his concern. Coiled up in one of the crotches of the large tree I was just standing next to, was a snake. My leg had probably only been a couple of feet from him as I worked. The snake was very obviously a poisonous viper of some sort.

The news about the snake traveled quickly and the guys around us got very excited at having that snake in our midst. The first idea that went around was to shoot it. But we vetoed that idea because a shot might alert the enemy to our position. So, another guy got the idea to kill it with his ax. Instead of getting close enough to strike the snake with a swing of the ax, he chose to throw it at the snake. He made a remarkably good throw. The ax spun through the air and the head hit the snake full. Once the snake was wounded, the soldier retrieved the ax and finished it off.

After all the excitement was over I whispered a silent prayer of thanks that the snake hadn't struck my leg as I worked right in front of him.

SNAKE IN THE BAMBOO:

We were on the move and passing through a large bamboo stand. There is lots of bamboo in the jungles of the Central Highlands.

This stand was one where you sometimes had to duck low to get through it. The trunks of all the bamboo shoots were growing in different directions and were all tangled together about four to six feet above the ground. Some stands of bamboo were bent over so low you had to crawl to get through. As you crawled, the branches always had a habit of reaching out and grabbing the equipment on your backpack. Exhausting. This wasn't a really bad patch because you just had to duck.

As I wound my way through the bamboo path, slightly hunched over, I saw the guy ahead of me in the column stop, look back toward me, and point upward. I nodded and he turned and continued walking. Since he was about 10 meters in front of me, I couldn't yet see what he pointed at.

As I approached the spot where he had stopped, I kept looking up into the tangle of bamboo. I was just about there when I finally saw what he wanted me to see.

Looped around the bamboo branches above the path was a lime green bamboo viper which was almost the same color as the bamboo. The bottom of the bent branches were about five feet above the ground, and he was about a foot above that. The snake was about three feet long.

A bamboo viper was supposed to be one of the deadliest snakes in Vietnam. It was one of the infamous 'two steppers', meaning that if you were bitten by one of them, you took two more steps and dropped dead. An exaggeration, of course, but it made the point as to how dangerous they were.

The snake didn't seem agitated, but I still skirted around it a bit so I didn't pass directly under it. Once passed, I stopped, looked back at the guy behind me, made sure he noticed me, and then pointed up at the snake. The guy nodded. Having done my duty, I turned and continued on through the rest of the bamboo.

Heaven only knows how many other times I passed close to one of those snakes without seeing it.

RTO ABOUT FACE:

I was an RTO in Charlie Company. We were in the Ia Drang Valley. We had been moving through the tall grass and sparse trees for a few hours. Joe, the company radio RTO, was walking in front of the CO. I was walking behind the CO, carrying the battalion radio. Things had been uneventful thus far that day.

Joe was walking about 10 to 15 meters behind the guy in front of him in our column. The CO was about the same distance behind Joe, and I was the same behind the CO. We always tried to keep the men of the company spread out as we moved through the landscape. Being spread out led to fewer casualties if we were ambushed.

I just happened to be looking down the column past the CO and Joe when I saw Joe stop in mid step, pivot around on his rear foot, and start walking back toward the CO. It was a perfect 'about face' move. Joe was a tall, good looking black guy. As he walked back toward us, his eyes were wide, white balls in his black face.

Surprised at his action, the CO and I both immediately stopped.

After he had taken about three steps toward us, he said one word, "Snake!" I looked past him and saw a snake on the ground coiled up in the middle of the trail. Apparently, it had tried to cross the trail between Joe and the guy in front of him, but Joe got there too soon. The snake quickly coiled up and was ready to strike. Joe saw it just as he was about to step on it.

Since the snake blocked our path, the CO grabbed the company radio's mike and called for everyone to stop. Soon, someone pulled out a .45, got the CO's permission, and dispatched the snake. We must have been sound compromised in some way already because normally we would not shoot a round without good reason. Shooting could have alerted any nearby enemy soldiers to our position, which we wouldn't want to do.

The guy that killed it picked it up to look at it. When he did, he squeezed its mouth open, and we all saw its very long fangs. We knew that meant it was poisonous. Joe was very lucky that he looked down just in time not to step on that snake. Once all the excitement was done, the dead snake was tossed to the side, and we continued on our way.

SNAKE IN THE ROCKS:

It was in September of 1967. I was working as an RTO in the Tactical Operations Center (TOC) at the 1/12th, 4ID firebase.

That morning we had just started to move the firebase to a new area of operation. The new site for the firebase had a 40 meter wide, B-52 bomb crater in the middle of it.

As we worked to set up the TOC near the crater, we had to weave our way around the dirt and large rocks that lay at the crater's edge. The ground was bare of most vegetation, because it was covered with several inches of dirt that had been blown out of the crater when the large bomb went off.

We had been working and walking a lot in this area for the past several hours. Dozens of us were involved in work nearby. Suddenly, someone yelled out "Snake!" We all looked down and there crawling on the ground among us was a two-to-three-foot-long bamboo viper.

Up to that moment, the snake must have spent all the time we were there, hidden under one of the large rocks that littered the ground. We all gave it a wide berth until some brave soul came up with a long stick and dispatched it (See Figure 34-1). All of us were grateful it had shown itself during the day instead of curling up with one of us that night as we lay sleeping.

BOA FOR THE GUYS:

We were in a patrol base in the jungles of the Central Highlands of Vietnam. We were scheduled to stay there three days. The following story was told to me by the guys that lived it that day.

On the second day in the base, a squad size patrol (eight to ten guys) was sent out to search the jungle area around our base.

The squad was out a couple of hours when one of the guys jumped over a log laying on the ground. As he did, he saw it move. It was a huge snake. He got the attention of the rest of the guys, and they came back to look at the giant snake. They all got pretty excited with the find. None had ever seen a snake that big before. There was lots of discussion about what they should do. They decided they should shoot it. So that's what they did.

After it was dead, one of the guys wanted to show their prize to the rest of the guys back at the patrol base. But he knew they couldn't carry the whole snake back to the patrol base—it was huge. Instead, he decided he would just bring the head back with him.

He started chopping the head off with his machete. After a few moments, he found doing that was a very hard job, much harder than he expected, but he finally finished.

Now, he had to decide how to carry the head back. They didn't have a bag or anything like that to put it in. His only idea was to stick his machete deep into the cut portion of the head and bring it back, skewered on the machete pointed upright in his hand.

He walked for over an hour with the giant snake's head on the end of his machete. But by then, his hand and arm were getting very tired. Finally, he said to himself, "Screw the guys back at the patrol base. I'll just tell them the story." He threw the snake head away and came back with this story.

All his buddies verified the story to be absolutely true.

Figure 34-1: This snake was spotted crawling among us as we were building a firebase. He was hidden for hours under some rocks, but eventually made a break for better cover. He didn't make it. This snake was considered very deadly. He was known as one of the 'two stepper' snakes in Vietnam.

Bamboo Viper—KIA

35. Top and the Smoke Round

Before I start this story, let me talk a little about our Charlie Company First Sergeant, or 'Top' as all First Sergeants are called. The First Sergeant is the most senior NCO in a company and almost always the right-hand man of the Company Commander. He's usually the guy that gets things done for the CO.

Our First Sergeant was a special man (See Figure 35-1). Top was loved by the soldiers of Charlie Company. He was the steadying force for us, all the time. He offered strength and guidance to everyone. His advice and counsel were highly valued by all, including and maybe especially, by the officers of the company.

I met him for the first time, briefly, the first evening I reported in to Charlie Company out in the jungles of Vietnam. Before I became the CO's RTO, I didn't have many opportunities to talk to him, but came to know him a lot better after I moved into the command group.

Top was a bear of a man. Husky to the point of almost looking fat, but he carried himself very well out in that jungle. He was a man's man. He was always very calm, but you just knew you shouldn't do anything to anger him.

As we traveled through the jungle, I remember sometimes struggling a bit to keep up with the soldier in front of me. Then, I'd see Top waiting ahead on the path I was walking. As I passed, he always asked how I was doing. He stood there and asked that of every man in my column. Often, an hour or so later, after I had been keeping a good pace while on the move, I'd look ahead and see Top waiting there on the path again. I had no idea how he caught up and then passed me during that hour, but there he was. That happened often as we walked through the jungle day after day. As our formation was walking straight ahead, he was zig zagging and walking faster than everyone else so he could keep an eye on his men.

After I transferred into the command group, he often walked in the same column I was in. I never saw the man trip. I was always catching my foot on some vine or root and almost going head over heels, but not Top. He was always steady and sure while on the move. He was quit a man.

This particular day, we were digging in for the night. The CO and the artillery forward observer (FO) were in a deep discussion as to our exact location. That information was needed before def-cons could be established.

Def-cons were 'defensive concentrations' or aiming points around our perimeter for our support artillery. Def-cons were almost always established late in the day when we were almost done building our perimeter. As they were being established, everyone was inside the perimeter.

The purpose for def-cons was to allow the artillery to quickly respond to any enemy attack on our position with accurate fire. Usually two def-con points were established around our perimeter. That way, if we were attacked, the FO could quickly radio back to the firebase and have them immediately begin shooting rounds at the def-con point closest to the enemy's position. The aiming point of the artillery could be adjusted as needed during the battle.

Once again, the first task to establish def-cons was accurately determining our own location in the jungle. Ground navigation in the Central Highlands was very challenging. Seeing landmarks was difficult because we were in triple canopy jungle, judging distance traveled was, at best, just an educated guess and the terrain maps were not always as accurate as they should be. Therefore, the start of setting def-cons could be very 'exciting', if we were not exactly sure about where we thought we were.

Once the CO, FO, and others decided on our most likely location, the second thing done was to bring in a smoke round. Smoke rounds were artillery shells that had three large metal pods inside them. The pods were filled with chemicals that smoked when the shell burst open in the air at a predetermined point in its flight. The three pods flew through the air where we could see the smoke, and then landed approximately where a normal high explosive (HE) round would land.

Smoke rounds were always used first because, as I said, sometimes we didn't know exactly where we were. Bringing in an HE round first could be very dangerous to us if we had guessed our location wrong.

The artillery unit has very accurate calculation methods to make their rounds land pretty much exactly where they want them to land. For def-cons, the main thing needed is an accurate location from the FO. Once that is known, they can do calculations to adjust the landing point of the rounds, in reference to the location of the FO. In other words, once they know where the FO is, they can adjust or 'walk' the rounds to different aiming points relative to the FO's position, by following the FO's instructions.

That process of establishing def-cons became very familiar to those of us who were usually within range of the FO's voice as he talked on the radio to the artillery battery. The whole process went something like this.

After the smoke round established the starting point, the FO would call for an HE round. Sometimes he would first adjust the landing point of the HE round. He adjusted the landing point by saying something like, "Add 100". That meant the HE round would land 100 meters further away. Or he could make it land closer to us by saying "Drop 50" or some other number of meters he chose. He could also adjust the landing point left or right of the previous round's landing point by saying, "Left 50" or "Right 100". The artillery battery made the calculations to make the adjusted landing point happen.

The FO would have predetermined the two def-con points that he wanted to establish near the firebase. I think they were established on two sides of the perimeter. He adjusted the landing points of the HE rounds by 'walking' them in to his desired def-con point, using the above process. Sometimes he could do it with only one or two targeting rounds fired, other times it might take three to five rounds. When a round landed where the FO wanted it to land, he'd tell the artillery battery to 'Mark it' as a particular def-con.

Once one def-con point was established, he would move on to the other with the same process, until both were marked, and ready to use if we went into battle at this night perimeter.

Usually, the FO would not establish def-con points too close to our perimeter. He didn't want to put us in danger from the targeting HE rounds. However, occasionally, he would misjudge the distance the previous HE round landed outside our perimeter. Then, in the opinion of those of us who overheard his instruction, the FO might over-adjust the drop amount. Meaning, we thought the last targeting HE round was 'pretty close', and that the 'drop' amount would bring the next round in way too close.

When that happened, we'd know to take cover behind a tree before the round was delivered, just in case. Many times, when that next round exploded, we would hear shrapnel buzzing through the air around us. That was often followed up by a few guys yelling something like, "Hey, too close!" at the FO. He would then say something like "Add 50 and mark it" to the artillery battery.

So, each evening, as we created our perimeter, part of our 'normal' life was to have high explosive artillery shells landing all around us.

That day, there was, apparently, a real disagreement about where we were. I think the FO won the discussion regarding what our supposed location was. He radioed back to the firebase and gave the coordinates of the point he wanted to have the first smoke round land. He thought those coordinates were significantly outside our perimeter.

He called for the round and looked in the direction he thought he'd see the smoke in the air. At that moment, I happened to be looking at our first sergeant (Top) leaning over his newly built hootch. Top was between the FO and me and about 20 meters away from me. He was adjusting a rope or something on his hooch.

Just then, I heard the burst of the smoke round in the air behind where the FO was standing. Three steel pods came hurtling into our perimeter. One of them, trailing red smoke, flew past Top's face by inches. The pod buried itself in the ground about five meters from his hootch and continued spewing smoke from the ground.

Top, cool as always, turned toward the FO and simply yelled very loudly, "Lieutenant!" That one word said all that needed to be said. If that pod had hit Top's head, he would have been killed instantly. If it had been an HE round, many of us may have been killed or wounded.

A very embarrassed FO, quickly, very sincerely, apologized to Top, and made a major adjustment to the coordinates for the next smoke round. The next round landed well outside the perimeter. Obviously, we were not where the FO thought we were that evening.

It wasn't unusual to have smoke pods land close to our perimeter, but that was the only time I can remember they landed inside the perimeter. Thankfully, it was only a close call for Top and no one was hurt by any of the three smoke pods that came into our perimeter that day. It did make for a good story for us to tell though. I'm pretty sure Top didn't let that young FO forget about the incident very easily.

Figure 35-1: The gentleman in the middle was Charlie Company's 1st Sgt or Top as every 1st Sgt is called. He was a man's man in every sense and very beloved and respected by all the men of Charlie Company.

Our 1st Sgt (Top)

36. Evening Meals

As an RTO, I was part of the company's command group. The command group consisted of the CO, his two RTO's, the First Sergeant, and several other NCO's and enlisted men who had different jobs within the group. There were probably eight of us in the command group.

Our main source of food in the jungles of the Central Highlands was C-Rations—they were probably 95 percent of our meals. Now, in my opinion, C-Rations were not good, but they were not bad either. Of the 12 different kinds of meals available to us, most guys probably found five or six that they thought were OK and the rest were just eatable. But there was one meal that almost no one liked—ham and lima beans. When it came to C-Rations, we made do.

However, the command group didn't always have to 'make do' for our suppers. Sometimes we ate very well. Whenever it was practical, Top and one of the other NCOs served as our cooks for a special meal. One of them was from Guam and I think the other was from Okinawa, so they were both very familiar with using rice in meals. Between them, they carried the rice, two or three medium sized pots, some cooking tools, and a few flavorings for the meals.

If we decided we had time for a group meal that day, before we started to work on our night perimeter we all searched through our C-Rations and found what we wanted to contribute to the evening meal that day. Our contributions were always one can (our pick) of the 12 meat choices from the various C-Ration meals. For example, the beef steak, the boned chicken, the turkey loaf, or the meatballs and beans would all be good choices.

After getting our one can contribution, while the rest of us continued our duties building the night perimeter, Top and the other NCO went about preparing the evening meal. Most of the time they would start by building a small cooking fire. Then they would open all the C-Ration meat cans collected from the members of the command group, and dump them all together into one of the cooking pots. There were potentially up to eight different cans of meat contributed. Almost every time they cooked, there was a different combination of meats. Therefore, each meal was a bit different from every other meal they cooked.

They put the pot of C-Ration meats on the fire and began heating it (See Figure 36-1). They put rice into the other pot and when the time was right, began cooking it. When all was done, they called a temporary halt to work, and we all sat down in the jungle for a surprisingly good meal. C-Ration mix over Rice—a significant improvement on just plain C-Rations.

The first couple of times they cooked for us, they thought it would be a good idea to add a little of, what I called, liquid fire to the C-Ration mix. Those Pacific islanders liked their hot sauce. However, it almost killed a young midwestern kid—me. A couple of the other guys felt the same, and we convinced them to add the sauce AFTER we had taken our portions. They thought that was pretty funny.

Most evenings we didn't have time for a group meal, but the ones we did share were special. It was just a small thing, but it helped make our lives out in that jungle a bit better. I still really like rice to this day.

Figure 36-1: Top and another NCO cooking a C-Ration meat mix along with some rice for a special supper meal for the company command group. When we had the time to make it, a meal like this was very appreciated.

Cooking C-Ration Supper

37. Short Round

We were pulling palace guard. 'Palace guard' is what we called it when we were assigned to guard the battalion's firebase. That duty was much easier than being out in the jungles of the Central Highlands. It seemed to us it was comparable to the way old time soldiers lived when they actually were guarding a palace.

When I first joined the company in the field, we humped through the jungle daily, without a break for a couple of weeks. Then the guys told me that we were going in for palace guard the next day. No one really explained what that was, so I assumed we were actually going to guard an old palace in the jungle. This was, after all, the exotic orient, so that made sense to me.

The next day as we broke out of the jungle and I saw the firebase in front of us, I was very disappointed. However, I was not dumb enough to ask anyone where the palace was. I managed to figure out what 'palace guard' was all by myself.

Life in the firebase was pretty easy. Soldiers in the company only had to do daily patrols and perform perimeter watch duties throughout the night. The daily patrols were usually only done by small elements of the company so the rest of the company got to use most of the time to rest.

Palace guard usually lasted about a week for each company and then that company rotated out again to do search and destroy missions in the jungle. As they were leaving, another company came in to replace them as firebase security and live the easy life for a while.

One thing about life at the firebase was not so attractive. The firebase was frequently the target of enemy mortar attacks.

One day, I was standing near the perimeter of the firebase talking to one of Charlie Company's platoon leaders. Suddenly, off in the distance came the dreaded puck, puck, puck sounds of enemy mortar rounds leaving their tubes.

He and I both immediately knew what it was and each of us took off running to where we thought we should be during the attack. I didn't know where he was going but I was heading to Charlie Company's command bunker. I was one of Charlie Company's commanding officer's RTOs and that's where I belonged during any attack.

I knew I had time to get there because there was usually at least 30 seconds from the time mortars left their tubes till the time they landed. As I was running, I could hear our mortar guys at the firebase were already returning fire. I was physically between the sound of the enemy mortars and our mortar crews, so I knew the mortar rounds they were shooting were pretty much going right over my head.

As I ran, I glanced over toward the mortar pits in the firebase. As I did, I saw one of the mortar rounds come out of the tube and start turning sideways as it slowly climbed into the air. A short round! I did a quick calculation in my head and decided that it could very well land near me.

I was still nowhere near making it back to the well-built command bunker. Time was getting close with that short round on the way. Just ahead of me I spotted a small foxhole that didn't have any overhead cover. At that moment, that didn't matter to me. It was a hole I could get into.

I figured that short round was just about to land, so I dove into that hole, headfirst like it was a swimming pool. Luckily, I was wearing my helmet because my head crashed very hard into the far wall as I dove in.

Shaking my head, I sat up in the corner and noticed the foxhole was already occupied. At the other end of the foxhole, which was only about six feet long and two to three feet wide, was a good friend of mine from 2nd platoon. I think I made a funny face at him about my dramatic entrance, and he gave me a wry smile.

We hunkered down in that small hole waiting for the mortar attack to end. When it did, we both climbed out. Most of the enemy rounds didn't land near us, so sitting in that foxhole wasn't too scary this time. I told him about the short round I saw, and we both wondered what happened to it. I didn't think I heard it explode anywhere near us.

I was told later that there is a safety mechanism on mortar rounds. Apparently when they come out of the tube normally, they spin like a bullet. They have to spin a certain number of times before they are armed to explode on impact. That safeguard must have prevented the short round I saw from exploding inside the perimeter of the firebase. That was nice to learn.

After the mortar attack, we sent patrols outside the firebase to make sure the enemy was not going to follow it up with a ground attack. They did not. It was a typical hit and run attack that happened quite often at the firebase. That is why we took the time and made the effort to always build substantial bunkers there.

38. Chaplain Crash

We were busy building our night perimeter on the crest of a steep hill in the jungles of the Central Highlands of Vietnam.

Work on the landing zone and the rest of the perimeter was going well. I was digging a foxhole when I overheard some guys say that the LZ would be done shortly.

Of necessity that day, the LZ was being cut on the side of the steep hill. This was not unusual. In an LZ like this, the choppers wouldn't land, but would just hover near the sloping ground while their contents were unloaded. Once unloaded, they would usually just corkscrew straight up till they were above the tree tops and then fly away.

Just then, through all the normal sounds associated with 'digging in', I heard the sound of a helicopter in the distance. The sound of that helicopter coming closer got my attention. I wondered why it was approaching our position if the LZ was not done yet.

The foxhole I was digging was only about 10 meters from the edge of the oblong, almost complete LZ, so I had a good view of it. As I looked toward the LZ, I saw the helicopter just above the tops of the trees at the far end of the LZ. It was slowly coming into the LZ. But the LZ wasn't finished! Obviously, there was some misunderstanding between the pilots and the company radio operator.

As the chopper began its descent into the LZ directly toward me, I clearly saw both pilots looking to their far left as they came closer to the ground. The problem was, there were long, large branches still extending into the LZ on their right, which apparently neither saw.

Sensing something bad was about to happen, I sought cover behind a big tree. Just as I got there, the awful sound of the rotor blades hitting the tree branches filled the air. The blades began to break and fly apart immediately. Peeking around the edge of the tree, I saw the helicopter first begin to rock and then fall to the ground. When it hit the steep ground, it began to roll down the hill amid all the fallen trees on the floor of the LZ.

It probably only rolled once or twice and then came to rest amidst a tangle of trees and branches. It was laying on its left side with the skids on the downhill side of the hill.

Naturally, many soldiers quickly made their way to the wreck to help the crew. I was one of them.

When I got there, the lone passenger on the helicopter was still in the interior of the helicopter, but he was conscious and looking out at the soldiers who were there to help. He was wide eyed and appeared to be all wet. His first words were "Don't anyone light a cigarette!" He was soaked in fuel.

Several of the guys gently helped him out of the wreckage because he complained of severe back pain. Luckily, the helicopter crew members were only slightly injured.

That lone passenger turned out to be the battalion's Protestant Chaplain. He was coming out to visit us that evening to conduct a service for all the Protestant guys in the company.

After we got the chaplain and the crew out of the helicopter, we finished the LZ. When it was done, a dust off (medical) chopper came in to take the chaplain and chopper crew back to base camp for treatment. We also got in the rest of our supplies before dark.

A visit by one of the battalion chaplains was a fairly regular occurrence when we were out in the boonies. Their services were always well attended.

I'm Catholic, so I have many memories of the masses I attended out in the jungles in the middle of a war zone. The priest always brought a portable table with him to serve as the alter and had a suitcase filled with the other items he needed to say mass. A couple of the guys would volunteer to help him set up and also serve as his altar boys (See Figure 38-1).

Our Catholic Chaplain had a small habit he always did at the very end of mass when everyone says the words, "Thanks be to God." During the traditional mass, as those words are said, the priest's hands are in front of him, and his palms are upturned. After those words are said, the mass is complete, and the priest normally just puts his hands to his side.

Our Chaplain added a bit to the end. Just as he said the word 'God' he would raise his hands up slightly and then slowly turn them over towards the table (his alter) and gently touch the table with both his hands. It was just his way of ending the mass. He did it every time. The memory of his habit has stayed with me throughout my life.

To this day, at the end of mass, when we say the words "Thanks be to God", I always put my palms up, gently turn them over, and touch the pew top. When my kids were growing up, they both noticed this habit. Today, when we go to mass together, they both look at me at the end of mass and do the same thing with me. They know why I do it and smile at me when we do it together.

Funny how seemingly unimportant things can stay with you forever.

Later that afternoon, the downed helicopter was stripped of its weapons and some other gear, and that equipment was all sent back to base camp. We were ordered to completely destroy the rest of the bird. They didn't want the enemy to scrounge anything of value from it.

Before we left the perimeter the next morning, the engineers set C-4 charges with delayed fuses all over the chopper. Two hours after leaving the perimeter, word went out to the company that the timed charges were about to go off and we should take cover. When the helicopter did explode, it was obvious that the engineers were a bit overzealous in the amount of C-4 they used. In spite of being two hours away, small pieces of that helicopter came falling down through the trees all around us. The helicopter was, obviously, completely destroyed.

Figure 38-1: This is how the Catholic chaplain said Mass in the jungle. All the chaplains would periodically come out to visit the troops, even in a one-night perimeter. The Catholic chaplain would bring a portable table for an alter and a suitcase full of the other items needed to say mass. Word went around the perimeter that mass would be said at a given time. I think almost all the Catholic soldiers showed up to attend.

Mass in the Jungle

39. March 17, 1967 — St. Patrick's Day

It was March 17, 1967. We were on a search and destroy mission in the Central Highlands of Vietnam.

The evening before, the first sergeant told me some helicopters were scheduled to come in the next morning before we started moving out for the day. He said I should get ready to be on one of them to go back to the 4ID base camp for three days of rest. He said one of the other guys in the command group could take my radio while I was in base camp.

That evening I transferred my radio to the other guy's backpack.

Sending someone back to base camp for a rest wasn't too common, but it happened. It was my first time going in for a rest.

Early the next morning when the choppers came in, I jumped on one and had a nice ride back to base camp. When I arrived at base camp, it took a while to get to Charlie Company's headquarters barracks to report in. The whole trip back probably took two hours.

When I went into the company office, the guys there were very sullen. I asked what was going on. They told me that Charlie Company had contact this morning and took some casualties. I asked who the casualties were. They said they still didn't know.

It was a hard time for me over the next couple of hours. During that time, I found that we had two KIA, still no names, and several wounded. Finally, about midafternoon I went back to the office again and learned that the two KIAs were both members of second platoon, my old platoon. I knew both and one was a very good friend of mine. Both died from small arms fire.

I left the office and just began walking around the company area—crying. I felt very guilty I wasn't with the company. I knew it probably wouldn't have made any difference concerning what happened, but I felt I still should have been there.

One of the guys who was killed was from a small town in Illinois. He always received the most mail of everyone in the company at mail-call. The whole town he was from must have been writing him.

The other soldier, my good friend, was from Texas. He and I started slowly as friends. I don't think he liked the new guy when I first arrived, but our friendship built as time went along. I felt a great loss when I learned he had been killed.

That evening I went back to the company office and told them I wanted to go back out to the company as soon as possible. They scheduled me on a supply helicopter going out to Charlie Company the next day.

The next afternoon, I flew into Charlie Company's night perimeter. When Top and the CO saw me come in, they were surprised because I wasn't due to return for two more days. I told them I had to come back. I'm sure they understood.

To this day, on St Patrick's Day, whenever we are out with people, I try to make a toast of remembrance for my two friends who died on St Patrick's Day in Vietnam in 1967.

40. Shallow Foxhole

In early 1967, Charlie Company had been assigned the job of securing and then helping build a new firebase for the 1/12th Infantry Battalion.

Early that morning, Charlie company was air assaulted into a rare, large clearing in the middle of the jungles of the Central Highlands of Vietnam. This was the clearing where the new firebase would be built. Our job was to secure the clearing and the jungle area close to it. After we arrived at the clearing and secured it, we began patrols of the immediate area to ensure the NVA were not close by.

As soon as the area was declared secure, the artillery, mortars, and the rest of the equipment and personnel who normally compose a firebase began to be airlifted in.

The first priority in building a new fire base is to protect the battalion Tactical Operations Center with many, many sandbags and to build bunkers for both the battalion and company command groups. This work was done by both battalion and company personnel working together.

At the same time, perimeter bunkers are constructed by

the infantry company guarding the firebase. In this case it was Charlie Company.

Since this was a firebase, our normal, quick to make, simple foxholes with logs for overhead cover were not good enough. Instead, very sturdy bunkers were built on the perimeter. These bunkers were foxholes augmented with heavy logs and many sandbags to provide overhead, side, and back cover for the foxholes. They were built to provide substantial protection from both mortar and small arms attacks.

All the digging, filling sandbags, cutting logs, etc. takes a lot of time and a lot of effort by everyone. I spent most of my day helping build Charlie Company's command bunker.

By late afternoon that day, most of the critical construction was completed. Soldiers who were permanently assigned to the firebase then started making sleeping bunkers for themselves. These were holes usually big enough to fit a couple of cots and, like the perimeter bunkers, were reinforced with heavy logs and sandbag roofs and sides. The purpose of the heavy bunker was to protect the guys sleeping inside in case of a nighttime mortar attack.

Just before dusk, word went out that there was hot chow available tonight. That didn't happen often, so it was a special treat. We took turns going to the newly constructed mess area to get some welcome hot food. It was almost dark when I took my turn to walk over there. I got a plate full of food, ate it there, and was just heading back to the Charlie Company command bunker area.

Suddenly, I stopped dead in my tracks. Off in the distance I plainly heard the puck, puck, puck sounds of enemy mortar shells leaving their tubes. Since I was not too far from the command bunker that I helped build, I headed for it at a dead run.

Too late. By the time I got there, maybe ten seconds later, it was already overflowing with guys. I had to quickly find somewhere else to protect me from the soon to be landing mortar shells. I looked around and saw what looked like the start of a sleeping bunker. It was still just a shallow hole in the ground, but it had heavy logs surrounding it on three sides.

I ran to it and dove in. I landed on five or six other guys who were already in it. As soon as I landed, I poked my head up to see if I was below ground. I was not. I was below the three sides with logs, but I was above ground level on the side without a log. Not good. Those mortar shells, more of which I could hear still leaving their tubes, would be landing any second.

Then, I noticed the fourth large log they intended to use on the bunker was about five feet away, laying almost perpendicular to the hole. I scrambled up and somehow dragged that big log to the edge of the hole and lay down upon the guys again. Adrenalin is a good thing. This time when I checked, I was below the level of all the logs.

Just then, the enemy mortars began landing all around us. Even before they began landing, I was amazed to hear our mortar crews already shooting back in the direction of the enemy mortars. Good job, guys.

The whole attack lasted maybe five to eight minutes. As it went on, all we could do in that hole was lay there listening to the hiss of the mortar rounds coming in and then exploding. Some were very close to us. Intermixed with the explosions of the landing mortars, was the distant sound of more enemy rounds leaving their tubes. They had this area zeroed in well and really intended to pound us.

As the mortars were landing all around us, one of the guys in the hole said, "Well, at least if we get one in the hole with us, we'll never know it." I think someone immediately thanked him for the positive thought.

When it finally ended, we untangled ourselves from each other and climbed out of that all too shallow foxhole.

The firebase must have taken 80 to 100 rounds that night. I think we, sadly, lost one or two KIA's and had several WIA's during the attack, but it certainly could have been much worse. Especially so for those of us in that shallow foxhole if a round had landed in there with us. But it did not, and for that we were all thankful.

41. LP and the Grenade

We were in an overnight perimeter. I was on radio watch. It was sometime in the middle of the night. A night that was pitch black in the triple canopy jungle.

I had two radios to monitor. One was the battalion radio, which was connected to battalion command at our firebase. The other was the company radio which was connected to our three Listening Posts (LPs).

Each night we stationed three LPs outside of our perimeter. LPs were usually four-man teams whose job was to sit quietly in the jungle, about 75 meters outside our perimeter, listening for enemy soldiers trying to sneak up on our position. They were our early warning system to warn us against a surprise attack.

My job during radio watch was to get a situation report (sit-rep) from each LP team every half hour. The sit-rep had two purposes. The first was to make sure that at least one member of the team was actually awake. It was very difficult to stay awake in the darkness and relative silence of the jungle after the exhausting days we always experienced. The second purpose was to ensure that they were still alive. We needed to know that the enemy had not crept up on them and quietly killed them in the dark.

Another part of my job was to regularly radio in a sit-rep to battalion to report all was still OK with Charlie Company.

The radios we used could only be in transmit mode or receive mode. They could not do both at the same time. A talk button was on the handpiece. It was also known as the squelch button. When you wanted to talk, you pressed the talk button. When you were getting a call, you did not press the talk button because, while pressing the talk button, you could not receive a call.

At the end of a send conversation, when the talk button was released, the receiving radio heard a short sound we called a squelch, something like 'Chhh'. The sending radio did not make any sound when the talk button was pressed or released.

The LPs set the volume on their radio very low. Usually, the LP radio operator also kept the listen end of handpiece (shaped like an old fashion telephone hand piece) right up against his ear to muffle the sound of an incoming call as best as possible.

When LPs were called for a sit-rep, they normally only responded with a short press of the talk button which produced the squelch sound heard on the company radio.

On the company radio, my call sign was '6 echo'. The first platoon LP's call sign was '1-6 echo', second platoon LP's call sign was '2-6 echo', etc. When I made my call (in a whisper) to the first platoon LP it went something like this: "1-6 echo, 1-6 echo, this is 6 echo. Sit-rep, over." If all was ok, the LP would answer with a single squelch, 'Chhh', which I heard loud and clear on my radio handpiece, which was also pressed to my ear when I talked. My radio's volume was also set way down so as not to compromise my position.

After checking with the first LP, I went down the line asking for sit-rep's from each of the other two LP's. Half an hour later, I repeated the process all over again. That process went on all night long, every night we were in the boonies.

If there was something wrong, the LP responded with two squelches 'Chhh, Chhh'. I usually responded with '1-6 echo, 6 echo, say again, over' to make sure that the double squelch wasn't just an accident. If the LP repeated the double squelch, I would ask a couple of quiet, yes / no questions to get an idea of what the problem was. One squelch response for yes, two for no. The first question was always, "Do you have movement?" meaning did the LP have sounds around him that might indicate enemy soldiers were moving nearby.

If the LP felt that there was movement or just needed to talk with the company commander for some reason, I'd quickly wake and brief the CO. Once the CO was on the radio, he'd try to further determine what the situation was with the LP and what action should be taken.

This particular night, I got two squelches back from one of the LPs and found they had movement and wanted to talk to the CO. I woke the CO and briefed him on the situation.

Once the CO got on the radio, the LP talked to him directly in a very quiet whisper. The LP thought they had movement nearby and asked for "Permission to throw a grenade." The CO told him to 'Wait one". A couple of other soldiers from our command group were awake by that time as well. The CO told them to go around the perimeter to warn each foxhole that a grenade was going to be thrown outside the perimeter. He didn't want anyone to be surprised, panic, and accidentally shoot toward our LP's.

Once all the foxholes had been warned, the CO checked

with the LP again and asked if he still thought it was necessary to throw a grenade. The LP replied in the affirmative. With that, the CO gave him permission to throw a grenade. Then he told the LP to sit tight after throwing the grenade to see what happened. The CO didn't want the LP to begin wildly running back toward the perimeter in the dark of the night after throwing the grenade.

In the silent blackness of the jungle, we all waited for the explosion of the grenade. The first sound we heard, however, was a hollow "Conk" as the grenade hit a tree. Then, three seconds later, the grenade exploded.

My first thought was that the grenade had hit a tree right in front of the LP and possibly bounced back right among them. That would be a disaster.

The CO immediately got on the radio and called the LP. We were all relieved when the LP answered. The CO asked if they were OK. They were. Then he asked what was happening around them. They responded "Nothing". It had apparently been a false alarm, like most LP incidents were. Word was sent around the perimeter that all was OK and back to normal.

We all then had a quiet chuckle at the incident and were grateful no one was hurt. A little later, everyone who was not on watch crawled back into their hooches to try to get a little more sleep before dawn arrived. After my watch was done, I woke my relief, briefed him on what had happened, and went back to my hooch to get some more sleep myself.

Soon, another day in the jungles of Vietnam would start.

42. Creatures for Doc

After a long day of trudging through the jungle, we had found a good spot to stop. We were all busy doing all the tasks required to make a secure night perimeter. I was working to build the command bunker.

The company had just left the firebase a couple of days before after providing several days of guard duty there (Palace Guard we called it). While we were at the firebase, our battalion surgeon must have gone around talking to the guys about a hobby he had. Doc told them he was collecting animals and insects from the jungles of the Central Highlands. Since doc was very well liked by all the men, the guys were keeping an eye out for anything doc could add to his collection.

The bunker I was building was fairly close to where our two radios were sitting on the ground. The CO was monitoring the radios and was also talking to the forward observer (FO) about setting up our nightly def-cons. Def-cons were pre-targeted artillery aiming points around our night position. They were plotted before nightfall so if we were attacked during the night, the FO could quickly get accurate artillery fire on the enemy, regardless of which direction he attacked.

As I worked, I saw the CO go over to the company radio and answer a call. He talked for a couple of minutes, then signed off. I didn't hear any of the conversation. Soon, he answered another call. This time, as he talked, I could see him smiling a bit. He signed off again. After he did, he came over to me and asked to borrow my hunting knife for a little while.

The hunting knife I had was a very high-quality Puma knife a good friend had given me before I left for Vietnam. My friend was a Vietnam vet himself and said he gave me that knife because when he was there, there were many occasions when a good knife would have come in handy. So, he wanted me to have one. Now, I guess it was coming in handy for the CO.

As I worked on the foxhole, I watched the CO using my knife to cut small branches off the bamboo near us. Then, he started cutting the small branches into ten to twelve inch lengths. He was working hard and soon he had a whole pile of the small sticks. I started to get very curious about what the heck he was doing.

Soon, I saw a soldier come walking up to the CO, very proudly carrying a large tortoise. Its shell must have been 12 to 15 inches across. He held it by the shell, while the CO gathered his sticks and started pushing them into the ground forming a circular corral, big enough to hold the tortoise. When the corral was completed, the CO told the soldier to set the tortoise down inside of it. Once inside the CO's corral, there was no escape for that tortoise.

Soon after that, another soldier walked up to the CO. He was very carefully carrying a C-Ration box. The CO, not being a foolish man, told the soldier to put the contents of the box inside his newly built animal corral. The soldier went over to the corral and very carefully opened one of the covers of the box and poured out the contents into the corral.

Out came the biggest, solid black scorpion I had ever seen. Easily as big as a man's hand. As soon as it hit the ground, it went scurrying about inside the corral, working its way around the tortoise, to check every part of the enclosure. The tortoise ignored the scorpion, and the scorpion ignored the tortoise. Both the soldier and the CO were delighted that the corral worked so well.

After the soldier left, I was one of the first to walk over to see the new Charlie Company mini-zoo. I asked the CO what this was all about. He told me that the battalion surgeon had asked the guys to collect any strange creatures they found. The two creatures in the corral were for Doc. He said this was just their holding pen until the first helicopter came into our LZ that night. Then we'd send them back to Doc at the firebase.

I went back to work. As I worked, I occasionally looked over at the CO. He was having himself a very good time, teasing that big scorpion with a long stick. I found myself hoping he wouldn't decide to adopt it as a pet and ask me to carry it around in a C-Ration box for him.

When the first chopper came in, the CO himself forced the scorpion back into the C-Ration box and then had that box put inside a bigger C-Ration box to double box it for safety. Then both the Scorpion and the tortoise were loaded onto the helicopter and sent back to Doc at the firebase. I'm not sure how Doc kept them captive there, but I'm sure he didn't have a fancy corral like we did at Charlie Company's night position.

43. Patrol Base Cookout

Normally, we moved to a new overnight position every day. However, sometimes the brass decided we should spend two, or more rarely, three nights in the same perimeter so we could thoroughly explore an area with squad or platoon size patrols. I think, sometimes the purpose was to just give us a little rest from our daily travels through the jungle. We learned this patrol base was to be our home for three nights. To us, that was very good to hear.

Over the first two days in the patrol base, we sent out multiple patrols and found no indication the NVA were anywhere near this area.

I don't remember the story behind why our First Sergeant (Top) did this, but that is not important. When Top found out we were staying there for three nights and that the area was secure, he got on the radio and pulled some strings back at base camp. I think he had this plan in place for a long time and thought this was the perfect opportunity to execute a very nice surprise for the men of Charlie Company.

The afternoon of the third day, several helicopters began

arriving in our LZ. Even the CO didn't know why they were coming in.

The helicopters were filled with all the fixings for a first-class jungle barbeque. There were large wire grill tops, charcoal, cooking tools, steaks, hot dogs, several side dishes, and even ice cream. The hot dogs were backups in case we didn't have enough steaks. All the food came out in the normal insulated food containers used by the Army for transporting fresh and cooked food, hot or cold. There was enough food for over 100 hungry men.

After it all arrived, Top and the other senior NCOs began directing the cookout. They had some of the guys clear out an area and then poured the charcoal on the ground. After starting the charcoal, they placed the portable grill tops over it. They organized a chow line to hand out the side dishes. Then they started cooking the steaks and hot dogs over the charcoal (See Figure 43-1). Delicious smelling smoke filled the air.

The CO and all the men were very impressed that Top could arrange something like this. Steaks? Ice cream? Wow!

When the first batch of steaks were close to being done, guys started going through the chow line getting potatoes, veggies, and other side dishes. Then they'd go over to where the steaks were being grilled and get a nice juicy steak added to their plate.

It was all great for everyone. The only food most of us had eaten for weeks were C-Rations, so this was an incredible feast. The ice cream for dessert was the topper (See Figure 43-2).

Once the first group was done eating, a few of them were selected to go outside the perimeter to relieve our OPs (security) so they could come in to eat as well.

There was one additional surprise for the CO that afternoon. It was an unplanned visit by the battalion commander, who arrived in the midst of the meal. The CO said later the battalion commander was not very happy, at first, because he was not told about this either. The CO managed to calm him down and assured him that security was still in place for the company and the danger level was low.

Once everyone had eaten, the grill tops, tools, insulated food containers, and the rest of the stuff were sent back to base camp on another batch of helicopters arranged by Top.

It was the only time something like this happened for Charlie Company all the while I was with the company. It was a special day and a special memory for me and everyone else. Thanks, Top.

Figure 43-1: Grilling up some goodies. The CO is on the far right. This cookout was a great memory for everyone in the company.

Patrol Base Cookout

Figure 43-2: What a treat. Ice Cream! It was even still kind of frozen when they started passing it out.

Ice Cream in the Jungle

44. FU Lizard

The second day of the patrol base, a squad size patrol (eight to ten men) was sent out to recon the area around the patrol base. While they were out there, they had an encounter with one of the more infamous creatures that inhabited the jungle we were in. They ended up with quite a story to tell when they got back to the perimeter.

First, let me describe this creature that shared the jungle with us. It was, what we soldiers affectionately called, a F--k You lizard. These lizards were large, maybe three to four feet long from the tip of their nose to the end of their long tails. We rarely saw them. But we regularly heard them.

In the evenings just before dark, the lizards would call out to each other from the treetops. Their calls were loud and frequent. The sound of the call sounded like they were saying "Huck que, huck que". We were American soldiers, so we translated this Vietnamese lizard's call to English, which made it "F--k you, F--k you". Hence the name F--k You lizard (I'll call it an FU lizard from here on). I always considered hearing their calls as part of the entertainment offered by the jungle.

Following is what that squad told us happened that day.

The patrol was moving quietly through the jungle. Everything was normal for the first few hours. They had found nothing. No indication of NVA presence anywhere. The patrol was nearly over, and they were not too far from the patrol base. Just then, one of the guys startled an FU lizard out from some thick undergrowth where it was hiding. It made a run for it and tried to hide again in a hollow at the base of a large tree. The soldier saw where it went and called his buddies to come have a look.

Unfortunately for the lizard, the hollow at the base of that tree was not deep enough to get its whole body into it. Maybe a few years earlier when it was smaller it fit, but not now. A good portion of its tail stuck out the opening of the hollow.

Naturally, the soldiers were all excited. First, about actually seeing one of these stealthy lizards, and second, because its tail was sticking out of that hole. They were almost back to the patrol base and had seen nothing at all indicating there were any enemy soldiers in the area, so they felt comfortable that they were not in any danger from the enemy.

Because they felt safe where they were, they began discussing what they should do about this rare opportunity concerning the lizard.

Some guys wanted to just leave it, but most wanted to get it out of the hole, kill it, and bring it back for the rest of the company to see. They knew a lot of the guys had never seen one. The guys who wanted to bring it back, won the discussion.

So, while one guy stood ready to shoot, two others grabbed the tail sticking out of the hole and began to pull. They soon discovered the lizard was STRONG. It apparently had a good grip on the edges of the hollow in the tree and they could not budge it, try as they might. After lots of tugging, they decided a new plan was needed.

After several suggestions, they settled on what seemed to be the best option. C-4! No, they were not going to blow it up, instead they would burn it out. C-4 is a very powerful explosive, but if lit with a match, it will burn, not explode. And it burns hot!

They packed a couple of small pieces of C-4 on the lizard's body and lit them. According to the guys, that got the lizard's attention real fast. It started wiggling around and then suddenly was backing out of the hollow. The guy who was waiting to shoot, did his job perfectly. Now they had a large, dead lizard on their hands, but they were still a little way from the patrol base.

One of the guys had a rope with him, so they tied it around the lizard's neck and began to drag it back to the guys. When they got back to the patrol base, they excitedly told everyone of their adventure. And they had the dead lizard as a trophy, to prove their story (See Figure 44-1).

Some of the NCO's were not happy about what they did—you know we're in a combat zone, right? But all in all, their reception was met with excitement by the rest of the guys in the company. Everyone was interested in what an FU lizard looked like and now we all knew.

Figure 44-1: KIA F__k You lizard. Brought in by one of our patrols. It got caught trying to hide in a tree hollow that was too small for it to get its tail into.

FU Lizard

45. First Roll of Film

A few days before, our previous CO who had commanded Charlie Company for the past six months had been replaced. His tour in Vietnam was over and he was going back home. Just before he left, I was told that I would be the RTO for the new CO of Charlie Company for only a week or so. Then I would be transferred to work at the battalion's firebase as the battalion commander's RTO.

I was looking forward to the transfer because I knew life at the firebase was going to be an improvement on my life in Charlie Company, a life that I had been living for the past eight months. I was also going to miss working for and with our previous CO. I had been his RTO for six months and we had grown to be friends. The new CO seemed OK, but it was not the same as with the previous CO.

During my last week with Charlie Company, something happened that had never happened before. A traveling PX came out to our patrol base. A traveling PX was just a bunch of foot lockers filled with stuff you'd find in the PX at base camp that you could buy. Toiletries are a good example of stuff they brought out. However, one item for sale intrigued me. It was a basic 35mm camera.

I had been using a Kodak Instamatic my whole time in Vietnam. It took pictures, but they were not real high quality. I used that camera because it was inexpensive and small enough to carry easily. I think by then I was on my third camera because the jungle and jostling of traveling had done in the first two after a few months using them.

I wouldn't even have considered purchasing that 35mm if I had been staying with Charlie Company. But, since I was transferring to the firebase, I felt I could take care of this more expensive camera there. So, I bought it and one or two 36 picture rolls of film for it. I don't remember how or why I had money (military script not real cash) with me, but I did.

After I bought the camera, a guy who had a camera like that back in the States, did his best to educate me on how much better this camera was than an Instamatic. Some of the advantages were better low light pictures, timed exposure settings, faster shutter speeds, and, as a bonus, sometimes you could get two or three more pictures from a roll of film than it was rated for. I was really excited to have that new camera.

I knew my time with Charlie Company and my life as an infantryman in the jungles of Vietnam was shortly coming to an end. I decided I was going to take a picture of everything that represented everyday life for us as infantrymen, so I wouldn't forget. Over the next couple of days, I did just that. I took pictures of Top cooking rice for us for supper, walking through bamboo patches, engineers packing C-4 around large trees to take them down for our LZ, digging foxholes, cutting overhead cover, everything I could think of.

A few days later, I had 33 or 34 classic pictures taken. Then

I was transferred to the firebase, never to re-join Charlie Company and the infantry life again. A week after that, I was sent to the 4ID base camp to get ready to leave on my R&R.

Once at base camp, I decided to finish that first roll of film so I could package it up and send it back home to get developed, which was my usual practice with the film I used. I took a few more pictures with the new camera in base camp and the small gage on top of the camera indicated 37 pictures taken. One bonus picture already. Nice.

I started turning the film advance lever and it stuck halfway through. Hmmm, that's funny. The camera's new, so maybe it needs just a little more pressure. I pushed a bit harder on the film advance lever. Suddenly, it advanced. Great. More bonus pictures. I took picture 38. Advanced to picture 39. Took it. Advanced to picture 40. No problem. Took 40. Advanced to 41. Oh oh. Something might not be right. There may be a problem here. I pressed the film rewind button and started to rewind. It didn't feel like anything was rewinding. I needed advice.

I found a guy who knew something about 35mm cameras. He told me there wouldn't be that many extra pictures on a roll of 36. I probably broke the film. He explained how there was now a roll of exposed pictures inside the camera that was not going back into the film canister. He said my only hope was to take the film out in a very dark place, put it in the plastic container the film canister came in originally, and send it in for processing that way.

That night, I got under my poncho liner in the darkest area I could find, opened the back of the camera, and, by feel, found the loose roll of film. I rolled it up and put it in the plastic container. I did all that in what I thought was pitch dark so I didn't ruin the film. I sent the container back to my folks with a note explaining the situation and said pay whatever you need to pay to get special handling for this roll of film.

Unfortunately, none of those pictures were successfully developed. I still miss having all the pictures I thought were so important to record infantry life in Vietnam. Some, I can still see in my mind's eye at least. But the others are, unfortunately, lost forever.

46. News on the plane

In July 1967 I had just been transferred from Charlie Company to the firebase to work as the battalion commander's RTO. Another part of my job there was to work in the TOC (tactical operations center) as an RTO as well. I worked at the firebase for about a week and then went on my R&R to Taipei, Taiwan. I went with a good friend from Charlie Company.

When the R&R was over, just after boarding the plane to take us back to Vietnam, one of the other soldiers onboard noticed the Fourth Infantry Division patches on the sleeves of our khakis. He stopped us and asked if we had heard the news about one of the units in the Fourth. We had not. He told us he had heard that one of the companies from the 1/12th was overrun and almost wiped out a couple of days before.

Shocked, we asked him which company. He said he didn't know. Both of us took our seats. I was sick to my stomach. Almost everyone I knew in Vietnam was in Charlie Company, both friends and acquaintances. What if it was Charlie Company that was overrun? We both asked all the guys around our seats if they had heard anything. No one had.

Charlie company was now entirely composed of replacements for the original soldiers that came over on the USNS Walker one year before. Replacements, like I was, had come in piecemeal over time. The replacements were for casualties, both KIA and WIA, and for guys that had been transferred out of Charlie Company to other units. All the original soldiers were no longer with the company because their one-year tour had ended, and they were back home. Despite that turnover, I still had several close buddies in Charlie Company, and many other guys I knew. Could they all be dead?

That plane trip was one of the longest of my life just thinking about the guys in Charlie Company. I should have been thinking of all the fun I just had on R&R but I couldn't. Charlie Company was all that I thought about.

When we finally landed in Cam Ranh Bay, we began asking everyone at the airport if they knew anything about the 1/12th. Finally, we found a guy from the Fourth. He told us the company that was overrun was Bravo Company.

Feelings of relief and gratitude came over me. Thank God it wasn't Charlie Company. I was so grateful. Then, guilt suddenly took the place of relief and gratitude. I felt guilty being glad it was Bravo Company. Bravo Company was still a company of brave US soldiers, many of whom were now wounded or dead. What the hell is wrong with me? But, on the rest of the trip back to the 4ID base camp, I came to grips with my feelings. I was still grateful that all my friends were OK, but I also grieved for the guys we lost in Bravo Company.

I also kept thinking about how the week before I had just been with the battalion commander, as his RTO, when he visited the Bravo Company CO. We landed in their newly cut LZ and walked through a lot of the men to where the company commander was. I remembered that I thought the men looked very tired that day, as they worked on getting their perimeter ready for the night. It made me very sad thinking that many of the young men I saw that day were now dead.

When I got back to the firebase, the guys there were still in shock over what had happened. One sergeant I knew was particularly upset. He told me the story of what happened, both to Bravo Company and at the firebase while Bravo was engaged with the enemy. Briefly this is the story as best as he could explain.

Bravo was out on a search and destroy mission in the Ia Drang Valley. As they moved through the area, their point element made contact with an unknown number of NVA. They immediately became pinned down. Bravo's CO sent his lead platoon to their aid. He then consolidated the rest of his company into defensive positions as best he could.

The relief platoon became pinned down as well. They were facing a vastly superior numerical force of enemy soldiers. Not knowing that yet, the CO sent another platoon to relieve the first one. Suddenly the entire company had enemy soldiers intermixed with them. They were being overrun and taking many casualties. It was later learned that Bravo Company was fighting a force of NVA regulars numbering about 1,000.

Charlie Company was relatively near to the battle site, so the battalion commander directed them to go to the relief of Bravo. About an hour or so later as they approached the battle area, the NVA broke off and withdrew. When Charlie Company arrived at the battle scene, it was terrible to see. Dead and wounded were everywhere, scattered throughout the tall grass.

An LZ was cut and helicopters were sent in to evacuate the wounded. Of the 100 or so soldiers of Bravo Company, 33 were killed and 37 were wounded. Charlie Company sustained five wounded. In addition, seven Bravo Company men were taken prisoner by the NVA. Two of them died in captivity, one of whom was from my hometown of Kenosha, WI. The other five were finally released six years later when all the Vietnam POWs were released by North Vietnam.

The sergeant told me that as the battle raged, the TOC at the firebase was the scene of much concern as they listened to the radio communications with Bravo Company. At the start of the battle, the Bravo company commander spoke directly with the battalion commander on the radio in the TOC giving an ongoing update on what was happening and asking for orders. Midway through the battle, the RTO that carried the radio (my job when I was with Charlie Company) took over the transmissions. He said the CO was dead.

The RTO then described the battlefield as best he could. He said the enemy were everywhere. Many guys around him were down. Suddenly, the radio went silent. The RTO was killed as well.

The sergeant telling me the story said that it was a terrible, helpless feeling listening to the desperation in the voice of the young RTO. Everyone present in the TOC was affected and in despair. The sergeant's emotion, even several days later, was still very deep. Though I was not present, I can still feel his despair to this day. I've always been grateful that I did not have to experience firsthand that terrible day for the Red Warriors.

47. Life at the Firebase

When I was in Charlie Company, we all looked forward to Palace Guard. Palace guard was when our company provided security for our battalion firebase. One of the infantry companies of the battalion always guarded the firebase while the other two conducted search and destroy missions in the jungles around the firebase. The S&D companies were almost always in the battalion's AO and within the protective range of the big guns at the firebase (See Figure 47-1).

The firebase was the home of the battalion's artillery and mortar support. The artillery was 105mm howitzers and the mortars were 4.2 inchers. These were the big guns that supported the battalion's infantry companies when they were in contact with the enemy. They also offered support to any other allied units that were in contact and within their range. The firebase was home to maybe 100 to 150 soldiers with various jobs and also the 100 or so infantry men guarding it (See Figure 47-2).

Daily life was easier for the infantry company at the firebase; however, the firebase was still no Holiday Inn. The main thing it offered was some needed rest. It was always irritating when we were unlucky enough to be palace guard when they moved the firebase. Helping move the firebase was a lot of work.

When a firebase was being moved, a small portion of the infantry company was always the last to leave the old fire base. Just prior to leaving, they helped destroy all the bunkers at the abandoned fire base. It was very dicey to be in the last group of helicopters to leave because if the NVA were around, that would be the ideal time for them to attack the small group of soldiers remaining. I always breathed a sigh of relief when the helicopter I was in was well away from the old firebase.

When we got to the new firebase, we had two to three days of very hard work helping build the new one. Very substantial bunkers for the perimeter were always built. These bunkers usually needed ten times as many sandbags to build compared to our simple overnight bunkers, which were just foxholes with a few sandbags supporting logs overhead.

Our perimeter firebase bunker was built like this. It started out as an 'L' shaped foxhole big enough for four soldiers to stand in shoulder to shoulder. The 'L' was sideways with the long side facing out toward the enemy and the short side on the left. The short side was dug with steps so you could easily get into the foxhole from the rear.

The hole was dug about three feet deep. Then one or two layers of sandbags were laid on the front edge of the hole. On both front corners of the foxhole, several layers of sandbags were piled up like columns. On top of those sandbags, a large log was laid parallel to the front of the foxhole. The opening formed was the front shooting port for the soldiers inside.

The right side and the back of the foxhole were built with more sandbags up to about the height of the log across the front with smaller shooting ports built in. The top of the back row of sandbags was finished with another big log at about the same height as the front log and running parallel to it. The rear entrance on the left side of the foxhole was protected with a wall of sandbags to its left. The overhead cover extended over the rear entrance of the foxhole.

Logs were laid from back to front over the top of the foxhole. Then one to two layers of sandbags were laid over the logs.

Before we started using hooches, the sleeping area for the team was usually behind the foxhole. The sleeping area was normally surrounded by logs or sandbags. The sleeping area was big enough to hold three sleeping soldiers. When we started using hooches, the hooches were sometimes surrounded by sandbags, too.

The logs or sandbags placed around the sleeping area served as initial protection for the sleepers in the event of a surprise mortar attack. Once awakened during an attack, we got into the bunker with its overhead cover which offered better protection from the mortars.

Building these substantial line bunkers took a lot of work and a lot of sandbags. But, in the end, they provided a great place for surviving a mortar attack or for fighting an attacking enemy.

Every day at the firebase, the infantry company sent out 10-to-25-man patrols to help ensure no enemy activity was in the area around the firebase. Most of the rest of the company got the opportunity to write letters, clean up a bit, and rest. Sometimes they even served hot chow at the firebase, which was a welcome change to our daily diet of C-Rations.

After I was transferred from Charlie Company to battalion headquarters group at the firebase, the firebase became my permanent home for my last three and a half months in Vietnam.

As a permanent resident, filling sandbags always occupied most of my day for the first two or three days after the firebase was moved to a new area of operations. The first thing we protected was the Tactical Operations Center (TOC). The TOC was composed of two rectangular trailer like units that each had a door on one of the narrow ends. The trailers were placed on the ground with their doors facing each other about three meters apart so there was an open space between them. Then, both trailers had sandbags completely covering the roof and the three walls without the doors. Logs were placed over the open space where the two doors faced each other, and sandbags were placed over the logs.

A sandbag wall was usually placed on both sides of the connecting space. That wall was built almost up to the roof of sandbags. The sandbag walls were built wider than the space between the units to leave just enough room between that wall and the walls of the trailers so a man could walk into the open space from either side.

Inside of the TOC was the radio equipment which allowed the battalion to communicate with the infantry companies, brigade, and division headquarters. Communication is vital in war. That is why the TOC was the first thing we protected (See Figure 47-3).

Once the TOC was done, we could build our sleeping bunkers. Our sleeping bunkers were built large enough to hold two cots with a little extra room for walking around. We always built them very sturdy with overhead cover so that we would be protected in a mortar attack. I always added shooting ports to my bunker so it could also serve as a 'last stand' in case the firebase was overrun (See Figure 47-4).

Most of the time while I was at the firebase, the guy who

shared my sleeping bunker with me was the Sergeant Major — the top NCO in the battalion. I ran in pretty high circles for a lowly SP4. The bad thing about sharing my sleeping bunker with him was that he, being a senior NCO, rarely physically helped with its construction. The good thing about it was, most of the time, he found a 'volunteer' or two, to help me with the work.

Once the firebase was completely built, daily life became much easier. I mostly worked in the TOC during the day and was also part of the nightly radio watch shifts.

Occasionally we got hot meals shipped out to us from base camp, but mostly we ate C-Rations (See Figure 47-5).

There were no shower facilities at the firebase. But there was more time available to take more or less regular steel pot 'baths'. We could do that because we usually had plenty of water available from the water trailers that we had at the firebase.

Latrine facilities at the firebase were very crude, but very necessary. Neither we nor the Army wanted 200+ men haphazardly relieving themselves just anywhere at the firebase. Since we occupied the firebase from seven to fourteen days, 'facilities' were always built, and we were expected to use them. The latrine facilities were usually built the first day we got to the new firebase by a detail of several guys.

For doing 'number one', we used cardboard tubes (open at each end) that once held artillery shells. First, a deep hole was dug, then partially refilled with the loose dirt. Just before the hole was completely filled, the cardboard tube was inserted into the hole and dirt was filled in around it to hold it in place. It was usually set in at an angle and stuck three feet or so above ground. We called these 'Piss tubes' and you did your business into the open end of the tube. The loose dirt in the bottom of the hole helped drain the urine into the soil. There were usually four or five of them scattered around the firebase, so you didn't have to walk too far to relieve yourself, no matter where you were in the firebase. Since the firebase was an all-male place, no privacy screens were around piss tubes.

For 'number two', one or two areas, usually toward the center of the firebase, but not too close to anything else, were chosen for slit trenches. Slit trenches are trenches dug in the soil, maybe two feet deep, 10 to 12 inches wide, and several feet long. The dirt from the trench was mounded up on one side of the trench. One or two entrenching tools were left stuck in the dirt for us to use. Sometimes, if there was no concealing vegetation around the trench, tent canvas was hung around it to give a little privacy.

To use the slit trench, you simply straddled it, pulled down your pants and underwear, squatted down, and did your business. You brought your own toilet paper that always came packed in C-Rations. After you were finished, you used one of the entrenching tools to cover what you left behind. Then you were done!

Rain made using the slit trench much more of an adventure because it was surrounded by slippery mud. Be careful to keep your balance!

This system worked pretty well. The only problem was the bugs that were attracted to both the piss tubes and the slit trench. The piss tubes always attracted small flying insects we called sweat bees. These were small bee-like looking bugs that were very slow. They were attracted both to sweat and to the urine at the piss tubes. They didn't sting so if they landed on your hand or arm, you could easily swat them and then thump them off your skin with a snap of the thumb and middle finger of the swatting hand. Nice and clean and very satisfying.

The slit trenches, naturally, attracted a lot of flies. It wasn't a pleasant place to be anyway, so you didn't spend enough time there to be bothered too much by the flies. When I covered my business with dirt, I always made sure anybody else's 'stuff' that was not completely covered, got a shovelful of dirt too.

When we prepared to move the firebase, the first thing that had to be done was to have an infantry company secure the new firebase location. That new location was usually high ground that was already a fairly open area. The fewer the trees, the quicker we could set up the new firebase. In most places, some natural open areas were scattered around, even in the Central Highlands.

The problem was, the NVA also knew the kind of areas we liked to choose for a firebase. Once when we were just moving into a new firebase, they mortared us the first day we were there. They had pre-plotted that area for their mortars just in case we ever used it for a new firebase. Almost all their mortar rounds landed within the perimeter. We had several casualties during that mortar attack. The NVA had done their homework well.

Once the new firebase area was secured, we began to break down the old firebase. First, the sandbags protecting the TOC were removed. Sometimes we took the time to empty them and bring the empties with us. But most times we just piled them up after removing them from around the TOC and used new sandbags at the next firebase. Antennas were taken down and packed. Equipment inside the TOC was secured for travel.

Once the TOC was ready for travel, cables were hooked up to each of the trailers. Then, a Chinook helicopter came in, hooked up to the cable on one of the trailers, lifted it up, and flew it to the new firebase. Waiting there were the soldiers who would start to make it operational again. The second trailer was moved shortly after the first. We always had enough men at the new firebase waiting to start filling sandbags to protect the TOC.

Then, the artillery and mortars were moved the same way, as well as all their ammunition. Security at both the old and new firebases was always a concern so the timing of when things moved was well thought out.

Most of the heavy equipment and most of the men were moved by Chinook helicopters (See Figure 47-6). Some of the men and lighter equipment were moved by Huey helicopters. It was a big operation and, most of the time, it went pretty smoothly. We did it enough so we got good at it.

As I said before, portions of the infantry company were the last to leave the old firebase after they collapsed all the bunkers. We destroyed the bunkers to prevent the NVA from using them for protection if they ever got caught coming in to examine our old firebase.

My jobs at the firebase were very enjoyable to me. When I first got there, I served as the battalion commander's RTO. In that job, I got to fly with him every time he flew, and he flew a lot. I got a great deal of flying time over the Central Highlands of Vietnam. That was fun, except, of course, there was always a chance that enemy soldiers below could shoot at us at any time. To my knowledge, though, that never happened while I was flying with the colonel. When I wasn't flying with the colonel, I worked part-time in the TOC.

After about three weeks of being the colonel's RTO, I was given a new job. They assigned me to work full time in the TOC. Another soldier was selected to be the colonel's new RTO. Working in the TOC was my full-time job for my last two and a half months in Vietnam.

In the TOC, I was the main RTO during the day. While working there, I was able to change the way things were done when I thought they could be done better or more efficiently. I also created all the schedules for radio watch for myself and the other guys that served as part time RTOs in the TOC. Overall, I found it was a very satisfying job.

Finally, living at a firebase meant that you were constantly exposed to loud noises. There were the constant sounds of the 105 howitzers shooting at all hours of the day and night (See Figure 47-7). The sound of the quad 50s, when we had them with us, occurred day and night as well. The loud sound of Chinooks constantly coming or going. And the occasional sound of a war plane of some kind, prop and jet, making a low pass over the firebase just for fun.

Occasionally, there was the sound of enemy mortar rounds exploding in or near the firebase, which was often followed up by the explosions of artillery and mortar rounds and bombs from planes delivered in retaliation for the mortar attack. Loud noise was a big part of life at a firebase.

Figure 47-1: A wide view of a firebase (small lighter area in center) with its Area of Operation (AO) surrounding it. This is just outside of the Central Highlands in the background. The battalion's infantry companies would search the area all around the firebase looking for the enemy. The companies could search up to seven miles away and still be within the protective range of the 105's at the firebase.

Firebase in the Central Highlands

Figure 47-2: Closer view of an entire firebase showing all the bunkers and gun emplacements associated with the battalion firebase.

Firebase

Figuire 47-3: The tactical operations center for battalion command. These two units held the communications equipment for the battalion. These units were the first thing protected with sandbags when a new firebase was being built.

Tactical Operations Center (TOC)

Figure 47-4: Me building a typical sleeping bunker at the fire-base. An entrance was located on the corner of the bunker, not shown on the right side of the picture. When the hole was done, logs were placed over the top of the hole and more sandbags were placed on the logs. Cots in the bunker provided much better sleeping accommodations than did an air mattress or just on the ground like when I was with Charlie Company.

Sleeping Bunker at Firebase

Figure 47-5: Hot Meal at Firebase Occasionally, hot meals were sent out to the firebase as a welcome treat for the men. Once or twice they had a mess tent actually at the firebase, but that was not the normal way things were done. I'm getting two plates of food here, one for the CO and another for myself.

Hot Meal at Firebase

Figure 47-6: This shows a Chinook lifting, what I think is, either a 105 or a water trailer. Can't tell for sure. This picture was taken during the dry season as can be seen by all the dust kicked up by the incredibly strong wind created by the Chinook. The guys closer to the Chinook sought protection from the stinging dust by hiding under ponchos.

Moving Firebase

Figure 47-7: A typical emplacement for a 105 Howitzer. Six of these made up a battery. The gun emplacements were scattered around the firebase.

105 Emplacement

48. Working in the TOC

In late June of 1967, the CO's 12-month tour was almost finished, and he was going to rotate back to the States. I had been his RTO for six months.

Over those six months we had become friends. Because of that friendship, he pulled some strings and arranged a transfer for me to work as the battalion commander's RTO. I would be working at the firebase.

He felt that working at the firebase during my last three and a half months in Vietnam would be somewhat safer than staying in Charlie Company. It would also be a somewhat easier life than humping the hills of the Central Highlands every day.

The transfer took place after I had spent a week working with the new Charlie Company CO and his new RTO's, teaching them how things were done in the field in Charlie Company. It was the first combat command for the new CO, so he appreciated any pointers I could give him. My old CO knew he would.

My new job as the battalion commander's RTO meant that whenever he went anywhere that needed radio communication,

I went with him. He flew in a helicopter over our area of operations (AO) almost daily. From the air, he directed the battalion's
companies as they were doing search and destroy missions in
the jungle. While he flew, he communicated with the companies via radio. I was there with the radio.

When I flew with the battalion CO (a lieutenant colonel),
I only took the radio, my M-16, ammo, grenades, and water. I
was not only his RTO, but also his unofficial bodyguard. Flying
every day was fun. Occasionally, we landed in one of the company overnight LZs, so he could talk to the company commander
face to face. When we visited Charlie company, I got to see a
few of my old buddies..

The colonel didn't fly all day long, so I also had other duties
to fill my time. The primary other duty was to work as an RTO
in the battalion's Tactical Operations Center (TOC).

The TOC was composed of two small trailer-like buildings
housing most of the communication equipment for the battalion. One trailer housed the radios and other equipment for Battalion S-3 (Operations). The other trailer housed the electronic
equipment for Battalion S-2 (Intelligence). I was assigned to
S-3 (See Figure 48-1).

The S-3 trailer contained a radio to talk to the companies
and another radio to talk to brigade. There was also equipment
to scramble our conversations with brigade. On the wall was a
topographical map of our AO upon which the current location
of each company was marked.

When I was carrying Charlie Company's battalion radio,
my call sign was '8-3 echo'. Whenever I wanted to contact battalion for the CO, I started the call with '3-3 this is 8-3 echo,
over'. The RTO on the battalion's radio in the TOC answered
with '8-3 echo this is 3-3, over'. Now, in my new job, I was the

RTO in the TOC talking to the companies in the field. I was '3-3'. I thought that was pretty cool.

I think I was the first former 'grunt' to work in the TOC since the division arrived in Vietnam. After a while, I felt there were some procedures that could work better if they were done in a different way. I felt it was important, so I made my suggestions to some of the officers and senior NCOs working in the command group. Most of the time, they agreed that my suggestions were a better way of doing things and my suggestions were implemented.

After almost a month of doing both jobs, I was told that I was not going to be the colonel's RTO anymore. I was going to be working full time in the TOC. That is what I did during my last two and a half months in Vietnam.

I continued to make suggestions and on my own changed the way some of the record keeping was done. Everyone continued to agree that most of the things I changed were better. It was very satisfying working in the TOC over that relatively short time.

Working at the firebase in the TOC was certainly an easier life than being in Charlie company. But it was not easy all the time. We moved the firebase to a new AO every one to two weeks. When we did, it was 18-hour workdays for three or four days while we tore down the old firebase and then built a new one. Everybody worked. There were literally thousands of sandbags to fill. Many were needed to protect the TOC. Then more were needed to build our own bunkers. It was hard work, but we all worked together and it usually went smoothly.

The TOC also needed around the clock RTOs on radio watch so, just like in Charlie company, sleeping through the night was still not an option. Several of us shared shifts monitoring the radios throughout the day and night. The one nice thing about being on radio watch in the TOC was that there was light inside. During those long night shifts, I could at least write letters home and, maybe, do a little reading. We occasionally got shipments of books from the Red Cross. The selection was limited but at least there were some books to choose from.

One of the pieces of equipment in the TOC was a scrambler for the radio we used to talk to Brigade. Every day, at a specific time, I don't remember when, we had to pull out a panel on the scrambler device and set the codes for the next 24 hours. The codes were set with a series of switches that we would move to a specific setting as prescribed by a code book. Each date had a specific code assigned and the scrambler for both radios, ours and Brigade's, had to be set with the same code for radio conversations to take place. Without both scramblers set the same, the receiving radio just got garbled sounds. The scramblers were used to prevent the NVA from monitoring our radio transmissions and perhaps gaining intelligence from that monitoring.

There was no scrambling for the radios we used to talk to the companies. Company radios (PRC-25s) were not equipped with a scrambler. It would also have been too impractical for the company RTOs out in the boonies to try to keep up with changing codes daily.

On the wall of the TOC was a large, topographical map of the area around the fire base. The surface of a topographical map is divided up into grid squares. Our maps had grid squares that were 1,000 meters by 1,000 meters. The map in the TOC was used to keep track of the current location of each of our companies out in the boonies. To accomplish that, periodic location sit-rep's were done between the companies and the TOC.

Three grid coordinates were chosen to serve as reference

points for the location function. The three grid coordinates were different each day. Each company commander had a code book that documented which grid coordinates would be used each day. He also had his copy of the same topo map that hung in the TOC.

An important part of the company CO's job was keeping track of his company's location. He did this using his topo map, visible landmarks, direction traveled, distance traveled, etc. When it came time for a location check, he would determine the company's current location on his map and plot it in relation to one of the three reference points for today.

I think the three reference points were always designated X, Y, & Z (Xray, Yankee, and Zulu). If 'Y' was the closest reference point to where we were, he would tell me to radio in a message something like this, "3-3, 8-3 echo, Our location, from Yankee, Uniform 1.2, Lima 3.4, how copy, over". That meant, from the grid coordinate defined as 'Y', we were up (north) 1.2 grid squares and left (west) 3.4 grid squares.

We did it like this to prevent any NVA listening to our radio transmission from determining our current location. The NVA would not know what the secret grid reference points were today.

When I was working at the TOC and got that message from one of the companies, I would take a grease pen and plot the location on the topo map on the wall using 'Y' as the reference point. I'd then mark that spot as the current location for that company. We did these location checks throughout the day for all the companies out in the field.

If any of the companies made contact during the day, the battalion commander would use this topo map to determine immediately the approximate current location of each of his companies. That knowledge would be a great help in determining how to best assist the company in contact.

Part of my job at the TOC was to accurately mark the map in the TOC with today's grid reference points. We used the same code book used by the company commanders, so we were always in sync.

One nice thing about working in the TOC that I really liked was that I always knew what was happening and, most of the time, why it was happening. It was the same as when I became an RTO in Charlie Company. As a grunt in Second Platoon, you rarely knew why we were changing direction or not stopping for the day yet. It was very frustrating just doing it without knowing why. But, as the RTO in Charlie Company, you got to know why decisions were being made. To me, that was so much better than just being told to "Do it" without any explanation.

Working at the TOC was very satisfying to me. I felt I made a positive difference in how things were run in the TOC and therefore, I think things were better when I left than when I got there. Just as I wanted it to be.

Figure 48-1: This is the inside of one of the two trailer units that comprised the TOC. This is the S-3 (Operations) side where I worked. The other trailer belonged to the S-2 (Army Intelligence) guys. The device on the far left is the battalion radio used to talk to the companies. The one to the right of me is the scrambler which scrambled our conversations with brigade. The device on the far right is the brigade radio. On the wall behind me is the map used to keep track of each company's current location.

Working in the TOC

49. Scorpion Sting

It was late summer, 1967.

Living out in the boonies at the firebase, meant that bugs of all types were just a part of life for us. We were all used to living with the bugs in the boonies.

One afternoon, I was standing and talking with three or four guys. Suddenly, I felt a pain in the meaty part of my left palm just below the thumb. I looked down and saw, hiding in a fold of my fatigue pants, a medium size scorpion. It was at the same place on my leg that my hand was as my arm hung by my side. Apparently, my hand had accidentally touched the scorpion and he then touched me back.

As soon as I saw it, I told the guys that I had just been stung by a scorpion, brushed it off my leg, and said I was going to see the doc. When I turned to leave, the guys were already surrounding the scorpion and preventing it from hiding anywhere. As I walked away, I heard one guy suggest building a fire around it to see if it would commit suicide like we heard in the movies.

As I walked to the battalion surgeon's bunker, pain was already radiating up the inside of my arm. I walked very deliberately and without panic because I was a veteran of Vietnam.

If I had been stung like this when I first came to Vietnam, I probably would have been in a full-blown panic. I would have been thinking only of my impending death. After all, in the movies scorpion stings were always followed by death. But now, I knew from experience that scorpion stings in Vietnam were usually not that bad, and that you could get treatment to lessen the effect on you.

When I reached the doctor's bunker, he was there, and I explained I had just been stung by a scorpion in the left hand. The doc went to a cabinet and came back with a small hypodermic needle. He gave me a shot in the left arm.

The shot didn't take the pain away, but the pain never got much worse. The next day, my palm was sore, but I was still alive, as I knew I would be.

I don't know what eventually happened to the scorpion. But I'm pretty sure he didn't make it.

50. Low Flyovers

The battalion firebase 1/12th, 4ID. The Air Force did a great job at supporting us grunts. Whenever we made contact with the enemy or were mortared by them, we could always count on the Air Force to help us out.

The forward air controller (FAC) was the guy that controlled the jets and A1E Sky Raiders (propellor driven fighters) that strafed and dropped bombs. The FAC flew in a small, single engine, Piper Cub like plane that could fly low and slow. Flying like that meant he could better see the battlefield below than could the bigger and faster planes.

The FAC was the person that we ground troops talked to when we needed air cover. After understanding what we needed, he then relayed instructions for targets to the war planes. It was a system that worked very well.

Sometimes the FAC would simply fly around, many times just above the treetops, to see if he could see any targets of opportunity. If he spotted enemy activity, he would call in the war planes to engage them. Sometimes he flew low and slow just to tempt the enemy to fire on him and therefore give away their position. If they were dumb enough to do that, hell would soon be after them in the form of war planes dropping bombs.

One time when I was the RTO for the Charlie Company CO, we saw a FAC fly over us at treetop level. We changed my radio to the FAC frequency and the CO contacted him to make sure he knew there were friendlies in the area. He did.

A few minutes later we could hear his motor off in the distance and then heard short bursts of automatic weapons fire from about the same direction and distance. It stopped, then it started again. Then stopped.

Concerned that the FAC was being shot at from the ground, perhaps without his knowledge, we again changed over to his frequency and the CO called him. When the FAC answered, the CO told him we heard some automatic weapons fire from his direction and asked if he was getting incoming fire.

The FAC laughed and said, "No incoming fire. It's just me raining a little death and destruction from the air with my M-16. All is good." We had a good chuckle at his response.

Occasionally, there were war planes on station in the area just waiting for calls to action, but none were coming in. I think just to relieve a little boredom, they sometimes did a low pass over the firebase just for fun and to entertain the troops on the ground (See Figure 50-1).

Once it was a VERY low pass. The jet came in so low; it just cleared the radio antennas we had sticking in the air from the middle of the firebase. A very strong blast of hot air immediately followed the low pass, blowing loose materials all around the area.

Another time, after a low fly-over by a jet, not to be outdone, the FAC did his own low fly-over (See Figure 50-2). That small plane with its whiney engine flying low and slow over us was a pretty funny contrast to the loud noise and wind of the jet preceding it.

Those FAC guys had a great sense of humor as well as a lot of courage. We grunts appreciated them a lot.

Figure 50-1: The Air Force pilots occasionally, just to have fun and to entertain the troops, did low flyovers of the firebase. Sometimes the jets did it, sometimes the propeller aircraft did it, and the FAC even did it a couple of times. They always succeeded in entertaining us. The Skyraiders were my favorite support aircraft. They could carry a huge load of bombs and flew much slower than jets. I always felt slower provided more accuracy, especially when they were dropping bombs close to us. No smart bombs back then, so good aim was important.

Low Flyover—A1E Skyraider

Figure 50-2: Even the FAC got into the act of doing a slow, low flyover. It was a comical change of pace from a flyover by one of the bigger, faster war birds.

Low Flyover—FAC

51. Puff and Flairs

One evening, enemy mortar rounds began landing inside of the perimeter of the firebase. Everyone took cover. I was in one of the substantial bunkers we had built for such occasions.

Even before the enemy rounds landed, our mortars began to return fire in the direction of the sound the mortar rounds make when they are leaving the tubes. The enemy attack lasted several minutes.

When it ended, there was concern that a ground attack might follow the mortar attack, so Puff was called in for support. Puff was the nickname for a C-47 gunship. That aircraft was armed with 7.62mm miniguns which could shoot up to 6,000 rounds per minute. It could also drop flairs that lit up the night.

When Puff arrived, the first thing they did was drop flares. The flares were attached to parachutes. They were great balls of light that slowly descend from the sky. Puff tried to keep at least two of them in the sky over us all the time. The flares were not bright enough to turn night into day, but they did help us clearly see out beyond our perimeter to ensure no enemy soldiers were coming.

The next thing Puff did was determine a target for their

guns. Since there were no enemy soldiers attacking us, they went after the bad guys that shot the mortar rounds at us. They were told the direction and estimated distance from the firebase that our guys guessed the mortar rounds originated from. Once that information was received, Puff began its fire mission (See Figure 51-1).

When Puff shot its miniguns at night, it was an incredible sight to see. From the ground at the firebase, we could see the small red identification lights on the aircraft circling above us and maybe hear its engines, but we usually couldn't see the aircraft itself because the night was so dark.

The entertainment began when it started shooting.

When they shot their miniguns, the show presented itself to us in three stages. The first stage was visual. From the ground, we could see the muzzle flash of the minigun, and the tracers being shot. Red streams of tracers undulated in waves through the sky as the plane rocked back and forth in the air. It was amazing because, although it looked like a solid stream of tracer bullets, the tracers were only every fifth bullet being fired. Awesome!

The second stage was presented as sound, and it depended on the direction Puff was shooting. Usually Puff shot parallel to or slightly in the general direction of the firebase. In that case, the crack sound of the bullets breaking the sound barrier followed the visual sighting of the bullets by one to three seconds. The crack sound was like hundreds of small firecrackers exploding at a rate of 6,000 per minute.

The third stage almost immediately followed the sound of the crack of the bullets. That was when the growl of the miniguns shooting the bullets reached us. The growl sound was almost always the same duration as the crack sound.

If the gunner shot a short burst, the sounds were 'Craaack', followed by 'Grrrrrr'. If, however, it was a long burst, the sounds were 'Craaaaaaaaaaaaaack' followed by 'Grrrrrrrrrrrrrrrrrr'.

If Puff was shooting in a direction away from the firebase, the growl sounds reached us before the crack sounds.

As long as there were no incoming mortars or ground attack on the firebase, Puff was a very enjoyable show for us grunts to watch. Plus, it was very comforting to know that firepower was there to help us if we needed it.

Figure 51-1: Shown here is a short time-lapse photo of Puff firing its mini guns at 6,000 rounds per minute. The line of bright lights are the muzzle flashes of the guns out the side of the plane. The line of tracers does not show how it looked in live action. In live action, the tracer rounds kind of undulated on the way down caused by the rocking of the plane. Growls and crack sounds also accompanied the light show. The clouds are lit up by the flares Puff had dropped a few minutes earlier.

Puff and Flares

52. Quad 50 & Twin 40s

At the firebase we always had an artillery battery (Five or Six 105's) and a 4.2mm mortar platoon (Three or Four tubes). A couple of times when the firebase was near a road system, we also had a Quad 50 unit mounted on a truck and a twin 40's self-propelled unit as part of our arsenal.

A quad 50 was a platform with four .50 caliber machine guns mounted together as a unit so they could all fire simultaneously. Each machine gun could fire 450 to 550 rounds per minute.

A twin 40 is like a small tank that had two 40 mm cannons mounted on it that could rapid fire together. Their rate of fire was about 240 rounds of cannon fire per minute.

Both of these weapons were awesome to watch in action.

I'm grateful I never had to see them in an actual combat situation, but there was one occasion when we got to see them test fire during the day.

On that occasion we were in an area that was primarily grasslands with sparse trees. The firebase was located on a low hill surrounded by the grasslands and trees. Word went around the firebase that a test fire of the quad 50's and twin 40's was scheduled for the afternoon. I made sure I was there to witness it.

The quad 50's started it off. First, the truck it was mounted on was moved to the edge of the perimeter so they would not be shooting over anyone's head. When all was ready, they began firing. It was incredible. With all four 50's firing together, it was like a water hose shooting out tracer rounds. And tracer rounds were only every fifth round that came out of each of the four barrels.

They were firing directly down the low hill, just above ground level simulating firing on an enemy attacking the fire-base. What I found most interesting was watching the .50 cal-iber tracer rounds hit trees and bushes. I was surprised when many of the rounds ricocheted off of them and then continued on in completely random directions. Some went straight up in the air, others went right or left. As the .50 caliber bullets hit the trees and bushes, the trees and bushes had no chance. They were cut down as sure as if a chainsaw were being used against them.

There was an incredibly loud roar as the 50's fired.

The 50's fired for maybe a minute or two and then stopped. Not a tree or bush in their line of fire was left untouched. Most-ly, just the splinters of their trunks were still standing.

Then the twin 40 was moved in and started to fire into the same area. As each 40 mm round hit, there was an explosion and a puff of black smoke.

The 40's had a rhythmic 'boom, boom, boom, boom' sound to them as they fired, which was echoed by the sounds of the cannon's shells exploding down range.

They also fired continuously for about a minute or two and then quit.

Although the twin 40 was quite impressive with direct, rapid fire of 40 mm cannon shells, I thought the quad 50 was more exciting to watch.

When it was all done, everyone agreed that any ground attack on our firebase would meet disaster with the help of those two awesome weapons.

53. H & I

The firebase had many functions. One of them was harassment and interdiction (H & I) fire. H & I was handled by artillery, mortars, and, if present, quad 50's. It was an ongoing activity, 24 hours a day.

H & I targets were picked by army intelligence. They would study maps of the areas around the firebase that were within range of the above weapons and pick likely trails or areas the enemy might use. Or they might receive human intelligence about enemy location or movement. The H & I targets were never supposed to be near any friendlies or civilians.

Once they picked the target areas, they decided what time of day and weapon was most appropriate for attacking these targets.

The reason for the H & I fire was accurately described by its name. Its purpose was to harass the enemy by randomly firing at them from afar and or interdict their planned activity or travel by the same random fire.

Since the targets were not a specific, known target, but a 'suspected' target, not a lot of ammunition was spent on each fire mission. For instance, a three to six round artillery volley might be considered appropriate. Fire mission times could be any time of day but were primarily done at night when the NVA did a lot of their movement.

The most impressive H & I fire to me was when the weapon of choice was the quad 50. A quad 50 is a platform with four .50 caliber machine guns mounted on it. It was designed so all four machine guns can fire simultaneously. Since the quad 50 was usually mounted on the back of a truck, we only had them with us a couple of times when our firebase was built near a road system.

Both artillery and mortars are considered to be indirect fire weapons. That is, they are not shot directly, point blank at the enemy, but are shot in arching trajectories. The trajectories and landing points of the artillery and mortar rounds can be very accurately calculated by their crews.

We were told that the trajectory and landing point of the .50 caliber rounds could be plotted just about as accurately as they could plot artillery rounds. When they wanted to attack a possible target that might be spread out over a larger area, the volume of rounds delivered by the quad 50 made it a good choice for that mission.

Since the quad 50 was designed to be an anti-aircraft weapon as well as an anti-personnel weapon, it could raise its muzzles straight into the sky to become an indirect fire weapon. When it was fired in the dark of night, it was an amazing sight (See Figure 53-1).

With all four 50's shooting at the same time, a column of tracer rounds raced into the sky. It almost looked like a searchlight. On a clear night you could sometimes see the tracers just disappear into the blackness. It was easy to imagine the terror inflicted on an unsuspecting group of enemy soldiers walking along a dark jungle trail when suddenly .50 caliber rounds started falling among and all around them.

For those of us that lived and worked at the firebase, H & I was a fact of life that you learned to live with. During the day, not so bad. You got used to artillery or mortar rounds being randomly shot off without warning throughout the day.

However, at night, you had to learn to sleep with the periodic firing of those weapons at any time and multiple times during the night. Sometimes your sleeping bunker was built very near to one of the artillery emplacements, which made the sound even louder.

But, after time, your sleeping mind simply learned to ignore the sound of a 105 howitzer going off within meters of your cot. On the other hand, if someone whispered your name to you while you slept, you would instantly be awake and alert and ready to go.

Such was life in a combat zone.

Figure 53-1: Harassment and interdiction fire in the dark by a Quad 50, which shoots four .50 caliber machine guns simultaneously. This short time exposure makes it look like a spotlight. In reality, it looks like a water hose spraying out tracer rounds. The sound is just as amazing.

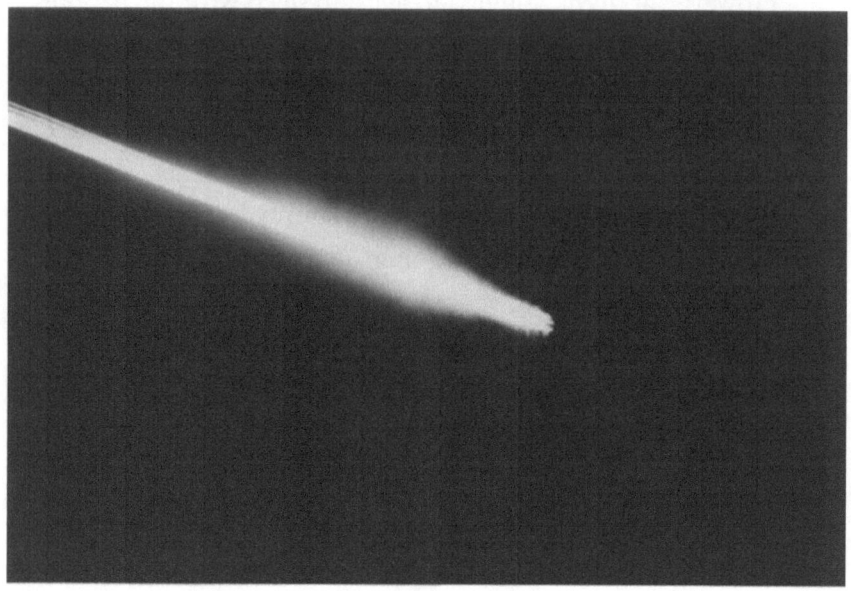

H & I—Quad 50

54. Muddy Convoy

In late summer, 1967, I was working at the firebase of the 1/12th, 4ID. We were in the process of moving the firebase to another area. This firebase was located in an area that had primitive roads nearby.

Normally, the firebase was moved from one location to another using only helicopters, both Chinooks and Hueys, because most of the terrain in our typical area of operation had no roads at all. This time, we were moving the firebase from an area near roads to another area with roads, so we were using trucks to do most of the move.

When I say 'roads', I'm being very generous. Most of the 'roads' were just single lane stretches of dirt. They were so rough that they were just barely passable even for large military vehicles. This day, the roads were in even worse shape than normal because we had had several days of heavy rain in the area.

Despite the condition of the roads and the muddy conditions throughout the firebase, the move was scheduled and the move would be made. Everyone had to do all their normal tasks for the move, in the mud.

Slipping and sliding as we worked was only one of the prob-

lems. Our jungle boots had cleated soles, so the cleats quickly filled up with sticky mud. They resembled reddish brown
stumps on our feet. Everything you put on the ground also got
muddy. Rain-soaked sandbags were much heavier than normal.
Handling them also left your hands covered with mud.

The artillery pieces were transported to the new firebases via
Chinook helicopters. Most of the rest of the stuff was loaded
aboard deuce and a half trucks. There were trucks everywhere
around the firebase. When they drove across the firebase, their
tires churned up even more mud.

Finally, everything was loaded. The infantry company guarding the firebase and most of the firebase personnel were loaded
onto trucks too. Twelve to fifteen soldiers were on each truck
(See Figure 54-1). There were also a few jeeps with .50 caliber
machine guns mounted on them mixed in between the trucks
for additional security.

I was on a truck about in the middle of the convoy. All the
vehicles traveling between the lead truck and the truck I was
on thoroughly destroyed the road before we got there. The road
became just a series of deep mud holes. As we slowly drove
through the mud, the truck rocked violently back and forth. I
was in a truck that was loaded half with equipment and half
with soldiers. We were sitting on top of the equipment, so the
rolling of the truck felt even more exaggerated.

Some parts of the single lane road became so impassable
with deep mud that the drivers deliberately drove off the road
onto the area just beside it to make a bypass around the mud
holes. We often crisscrossed the road, first driving on the left
side, then on the right, wherever it looked like the best route. It
was very slow going (See Figure 54-2).

Eventually, we reached an area that had a road that was

semi-maintained, and the mud wasn't so deep. It was also an area that had a few Montagnard villages along the way. We often passed pedestrians walking along the road. One of the things we saw a lot always got a few comments from the guys.

If there was a man and woman walking together, the man was always several paces ahead of the woman and the woman was almost always carrying a backpack full of something like firewood or food. Papa San never carried, only Mama San carried stuff. Invariably, some guy would say something like, "I gotta teach my girlfriend to do that when I get home," and we'd all laugh. Man humor at its best (See Figure 54-3).

When we finally reached the new firebase's location, unloading was just as messy as loading. It didn't take long for us to turn the formerly grassy area into a mud pit.

During my time in Vietnam, I was on several convoys but the muddy one was by far the most memorable.

Figure 54-1: Moving a firebase by truck. The troops are all loaded, and the .50 cal jeep is ready to blend into the line of trucks. The road ahead was mud filled and rough.

Troops in Trucks

Figure 54-2: Moving a firebase by truck. The 'normal' road was often too mud filled, even for these army truck beasts. In many sections, a new side spur had to be improvised and used while on the move.

Muddy Road

Figure 54-3: Many times, while on convoys, Vietnamese were seen walking along the road. If it was a man and a woman, invariably, the woman would have a load of something on her back and the man would be carrying—nothing—and always walking several paces in front of her.

Papa san & Mama san

55. Shot in the Arms

In my youth I had always been susceptible to strep throat. I seemed to catch it once a year before going into the service. Somehow, I caught it while in Vietnam and got very sick. Luckily, our battalion surgeon's bunker was just a short distance from my bunker.

When I started feeling sick, I went to see him. He quickly diagnosed strep throat. He gave me a penicillin shot and told me to stay in bed. I was in bed for several days, fairly sick, and he stopped in a couple of times to check on me. He gave me a second penicillin shot. After about five days, I was finally starting to feel a little bit better.

That day, he stopped by my bunker to see how I was doing. At that time, I was sharing my sleeping bunker with the Sgt Major, the top NCO in the battalion. The Sgt Major was in the bunker when Doc stopped by. I told Doc I was feeling a lot better. He said he would give me one more penicillin shot and that should do it.

He had given me the previous shots in my butt, so I pulled down my fatigue pants and underwear. He gave me the shot and then started talking to the Sgt Major. I pulled up my underwear and was just reaching down for my pants when I suddenly started to feel dizzy.

I was just starting to say, "Hey Doc, is this supposed to make me feel ......' when my legs gave out. I sat down on the cot and tipped over on my left side. I was laying with my left arm sticking straight out away from the cot and my feet were hanging down the side of the cot touching the floor. My left arm immediately started convulsing and involuntarily slapping my face over and over. When that began, I heard the doctor say to the Sgt Major, "Hold on to him! I'll be right back" and he ran out of the bunker. I started wondering why I was hitting myself in the face.

I felt the Sgt Major sit down on the cot next to my butt. He grabbed my convulsing left arm and stopped it from hitting my face. He also was pushing down on my right arm, apparently to keep me from flopping around on the cot. My whole body was convulsing.

The Sgt Major started talking to me, saying things like, "It will be OK, Doc will be back soon, just relax." As he talked, his voice became more and more strange to me. It started sounding like a man's voice on a recording that was playing too slow. I heard "Doooon't wooooorry, iiiit wiiiiill beee oooookaaaaay." Weird.

Then my eyes even started to get into the convulsion act. They began to move, very fast, side to side and things got all blurry.

Strangely, as all this was going on, I was not afraid in any way. I was just marveling at all the strange things happening.

Soon, I noticed some motion pass behind the Sgt Major. It was Doc coming back into the bunker.

I felt him trying to pull down my underwear, but I was

laying on them. He quickly gave up and pushed up the sleeve of my undershirt on my right arm and gave me a shot in that arm to, hopefully, stop the convulsions. Just after he did, the Sgt Major began massaging the arm where I got the shot. That wasn't helping me because I thought, "He's hurting both my right arms." In my head, I actually believed I had two right arms because I could feel pain in both of them.

Again, I heard him talking that strange, slow motion, talk saying "Theeeeeere, dooooose thaaaaat feeeeeeeel beeeeeeettt-teeer?" It didn't and I told him so. But he didn't stop rubbing my right arms.

Soon, my eyes stopped convulsing and I could see a little better. I was looking past the Sgt Major and I saw a shiny light on the wall of the bunker. I knew that there was not a shooting hole there, so I wondered what that light was. I suddenly got very excited when I realized, "I can see around corners! That's the shooting hole behind him." It made perfect sense to me. The shiny thing I actually saw was something metal hanging from the wall.

All the while this was happening, I was talking my head off describing what was going on. Soon, my ears started clearing up a bit and I heard myself talking. I sounded like someone trying to talk with an overdose of Novocain. Absolute mush sounds were coming out of my mouth.

Then I noticed the Sgt Major's stripes on his sleeve (See Figure 55-1). They looked, to my eyes, like a thick hairy mass, not the neat black camo strips they were. I heard a voice from the foot of my cot and looked up to see the Doc standing there. But he didn't look like the Doc.

The Doc was a normal looking guy with normal eyebrows and a small black mustache. The gentleman my eyes saw at the foot of my bed, had super thick, black eyebrows that were connected to one another—you know—a unibrow. He also had an incredibly thick Fu Manchu mustache that totally filled up the space between his nose and upper lip and went down the sides of his mouth all the way to his chin.

I began laughing. He looked so funny to me. Then, my ears began to clear up some more and I heard my laugh. It sounded like I was gagging on something, not laughing.

Finally, everything started to get back to normal. Doc began to look like his old self, and I could actually hear their voices clearly. I managed to tell the Sgt Major to stop rubbing, what was again, my only right arm. He did and he also stopped holding me down because the convulsions had stopped.

The whole episode probably only lasted five minutes, but, wow, what an intense five minutes. Again, I was not afraid or concerned at all during the whole time. Don't know why.

Doc told me to never take penicillin again because I had developed a severe allergy to it. He said it could kill me if I took it again and it was lucky it didn't kill me this time.

There were lots of ways to die in Vietnam. I've always thought this would have been a very sad way to go.

Figure 55-1: Our battalion Sgt Major (the top NCO in our battalion). He was my bunker mate for a while. Note the stripes on his arm — for a while, to me, these didn't look quite like this.

Sgt. Major

56. Susie

I had been transferred to the battalion firebase from Charlie Company just a few weeks earlier and was just getting comfortable with my new responsibilities. Around that time, someone, I don't know who, picked up a stray, maybe three-month-old, brown puppy, and brought her out to the firebase.

I had grown up with dogs all my life and loved them. As soon as I saw her, I knew I had to make sure she was taken care of. I can't remember who gave her her name, but we called her Susie.

Susie adapted to life at the firebase very quickly. Even with all the noise. Throughout the day and night, when the 105 howitzers had fire missions, she had no trouble being around them as they fired their volleys.

She would spend her days roaming around the firebase saying hello to everyone who was there. Everyone always gave her a bit of food as she said hello. She became a regular camp dog. Sometimes she entertained herself and us by chasing big bugs around.

Whenever we moved the firebase, I made sure I grabbed Susie and held her in my arms as we choppered to the next firebase location (See Figure 56-1). She even got used to flying in a noisy Chinook after a while. I also took the time to teach Susie some of the tricks all dogs should know how to do. It was great fun having a dog around (See Figure 56-2).

On the day I was packing up to leave the firebase for my trip home, I made sure I had my picture taken with her. It was the last picture taken of me in the boonies of Vietnam.

After I got back home, I often wondered what happened to Susie. At one of the 1/12th reunions many years later, someone shared an article with me, written about Susie. It was published in the Ivy Leaf newspaper sometime after I left Vietnam. She had become somewhat of a celebrity as the mascot for the 1/12th in Vietnam and, at that time, was still doing well.

Figure 56-1: Another firebase move. This time the troops are going by Chinook. This was Susie's first Chinook ride. Chinooks are very noisy, so I tried to protect her ears as we boarded and also held her tightly to reassure her that she was OK.

Susie Boarding Chinook

Figure 56-2: This picture was taken very near the end of my tour. I spent quite a bit of time teaching Susie basic dog tricks so she could better entertain the troops around her at the fire-base. I had fun teaching.

Susie Learning Tricks

57. USO Girls

In early October of 1967, I was a short timer. I had about two weeks left of my one-year tour in Vietnam.

I was working at my job as an RTO in the Tactical Operations Center (TOC) at the firebase, when word came around that we were getting VIP visitors today. Some members of a USO show that was playing at the 4th Division base camp were coming out to the firebase to visit the troops.

I was on radio watch when they flew in so I couldn't go out and see their arrival. Just after they got there, one of my buddies was nice enough to stop by the TOC to excitedly tell me that the visitors were four beautiful girls and three or four guys. They were being given a tour of the fire base by some of our senior officers.

About an hour after they arrived, my shift ended. I left the TOC to see if I could get a peek at the visitors. Actually, all I wanted was a good look at the girls. There was a large group of soldiers standing nearby, and I assumed that was where they were. I wandered over that way, and I soon spotted the young ladies. They were almost surrounded by soldiers.

All the USO people, including the girls, were wearing fatigues so they wouldn't stand out if Charlie was spying on the firebase. However, the four girls had makeup on, and all had their hair done up nicely. And, yes, all four of them were very beautiful, even in fatigues.

The USO people were trying to set up a picture of them with all the soldiers standing around them. Since there were so many soldiers already, I decided my short, distant view was going to have to be good enough, and I started walking away.

As I walked, suddenly one of the girls yelled out to me to come over and join the group. I was surprised, but it didn't take me even a second to change direction and head toward the group.

They were all gathered as for a typical group picture. The four girls were standing in the middle of the front row and the soldiers were crowded on either side and behind them. I had no idea if any of the guys who were part of the USO show were in the group or not. But I didn't care.

As I approached the group, I thought I'd just fit myself into the back rows somewhere. My new friend had a different idea. She waved to me and said, "Come right here." 'Right here' was directly in front of her! Then she said, "Just kneel down." So, I knelt in front of her, facing the camera.

As I did, she put both her hands on my shoulders and squeezed just a little. I still remember the feeling of her hands. Then, the smell of her perfume absolutely enveloped me. I'm pretty sure, if I hadn't been keeling already, my knees would have buckled. What a wonderful scent that was.

An eternity later (probably only a minute), the picture was taken. I stood up, shook her hand, thanked her, and walked away. I've always amused myself by imagining that she liked me—a lot. If only circumstances had been different, we probably would have gotten married and had a long happy life together. Right!

I'd been in Vietnam and out in that jungle for almost a year. Only a five-day R&R got me away from that environment over all that time. Smelling the perfume and having a pretty girl touching my shoulders was an overwhelming feeling to me. I guess, until that moment, I never realized just how much I missed home and how much I looked forward to going back.

That picture with the USO girls turned out to be an experience I would never forget. Wish I had a copy.

58. Extra Memories

RAIN

The rainy season started gradually. It began with brief daily showers occurring about mid-morning and again about mid-afternoon. Then, day by day, it seemed the brief showers began to last longer and longer until the morning rain blended in with the afternoon rain. Then it gradually began to rain almost continuously throughout the day.

Almost nothing you owned was dry. The precious items that you wanted desperately to keep dry, like stationery, were kept tightly wrapped in your rolled up poncho. But some of the time, even that didn't work so well.

We usually didn't wear our ponchos as raingear, especially when we were on the move. Some guys wore them while they were in the firebase, but I never did.

Fortunately, our nylon poncho liners still kept you somewhat warm even when they were completely wet. When you were soaked from the rain and got chilled to the bone in the evening when it got cool, you could wrap yourself in your poncho liner to get a bit warmer.

The rainy season lasted a couple of months. Then the rain

gradually tapered off and, for the most part, the weather was mostly dry with only occasional brief showers the rest of the year (See Figure 58-1).

4S & 5S

We called the beer and soda that the Army regularly supplied us with, 4s and 5s. That was the radio code for them. When they were shipped out to us in the field, usually with our C-Rations, the NCO's distributed them. Our normal ration was two cans of beer and three cans of soda each. Various brands of beer and soda were part of each shipment. I don't remember the beer brands, but the soda included Coke, RC Cola, and, I believe, 7-Up.

I didn't care much for beer at that time, especially beer at 85 — 90 degrees or higher, so I almost always traded my beer for soda. That was easy to do because some guys really liked their beer. I especially liked RC Cola because, to me, it was the best soda to drink warm. I thought warm Coke just exploded with foam in your mouth when you drank it at those warm temperatures.

The negative to having five cans of beer or soda was that you had to carry them. At 12 oz each, they added up. Usually, if you were going to move the next day, you saved a couple cans for the move, and drank the rest during the evening.

HAIRCUTS

I usually kept my hair pretty short while in Vietnam. Most infantry guys did. Short hair was easier to care for out in the jungle and there was less territory for bugs to hide in. We often got our haircuts while we were in a patrol base or at the firebase. One or two guys in the company volunteered to be our barbers. They generally had some barbering experience, but it didn't matter really. We never worried about looking good out there.

The barber guys carried hand clippers and scissors for doing their work. The haircuts generally only took a few minutes to 'Shorten it up', so they could get through most of the guys looking for a haircut in not too much time (See Figure 58-2).

SMOKE GRENADES

All of us carried smoke grenades. As an RTO, I usually carried four of them. The grenades came in red, green, purple, and yellow smoke. We used smoke grenades to communicate with aircraft. When we wanted the aircraft to know exactly where we were, we told them, via radio, that we were 'throwing smoke'. We'd pick a random color smoke grenade to throw. Then we asked the aircraft to identify the color of the smoke. We did this in case the NVA were monitoring our radio transmissions. The NVA might also throw a smoke grenade (probably captured somehow), to try to lure the aircraft into shooting range, but that didn't happen very often. If the pilot identified the color we threw, he'd know our location correctly. If the pilot identified a different color smoke than we threw, that alerted him to avoid that area.

FLYING

I spent a lot of time in helicopters while in Vietnam. They were the major way we were transported around in the boonies. Huey's could carry seven or eight infantrymen at a time. We'd always traveled with the doors wide open. No seatbelts were ever worn.

We would all squeeze onboard. Three or four guys would sit on the passenger seats behind the pilots, a couple more sat on the floor between the seats and pilots, and a couple guys might sit in the door openings dangling their legs out.

If the helicopter made a hard turn, the body of the aircraft might be almost sideways in the air. Even though you might be looking out the door and see only the ground straight below, there was never a worry about falling out. Centrifugal force kept you tightly in your seat or on the floor. That was fun to do. Too bad the concern about where you were going tempered the fun a bit (See Figure 58-3).

I think sometimes the pilots liked to mess with their passengers' minds a bit. Often, when flying in a formation of three, I thought they flew much too close to each other (See Figure 58-4). After all, they had the whole damn sky to fly in, why did they have to almost touch rotor blades as they flew along?

One time, the chopper I was in was flying very low level and fast in an area that was mostly grasslands with sparsely growing trees. As we followed the earth's contours, we approached a large, but not very steep hill. At the top of the hill were two tall trees. I was looking to the front as we headed up toward the top of the hill. I just knew that pilot had decided to fly between the trees, I'm sure just for fun. Just before going between them, though, he kind of tipped the helicopter slightly sideways. I think maybe he thought there might not be quite enough room to fit between them flying the normal way. Thanks buddy. That was fun—not.

AIR ASSAULT

Occasionally we did what we called an 'Air assault' also known as a 'Combat assault'. That meant that either the whole company or an element of our company was taken via helicopter to a new area and inserted on the ground. Most of the time this was done to surprise the enemy by our quick arrival. Many times, the area for our next firebase was secured by using an air assault to get the initial troops in.

Usually for an air assault the Huey's flew in groups of three (See Figure 58-5). They all tried to land in the LZ at the same time. As soon as they touched down, or perhaps a little before, we'd jump out of them quickly and head for cover on the edge of the LZ (See Figure 58-6).

Most of the time, there were no enemy soldiers nearby. Occasionally, though, enemy soldiers were there waiting for the choppers to land and took them under fire just before or just after they landed. When that happened, the LZ was known as a 'hot LZ'. I was lucky and I never flew into a hot LZ.

Figure 58-1: These are actually two pictures. The upper is a picture of part of the firebase during a deluge. Usually, the rain was just slow and steady for hours on end. This time, it was a cloudburst. The lower picture shows the results of the rain. These two NCO's are bailing out their bunker after it almost filled with water. My bunker was bone dry, and it was fun mentioning that to them.

Heavy Rain & Bailing out Bunker

Figure 58-2: Most of us wanted to keep our hair short while out in the boonies - easier to care for. These are four guys from Charlie Company at a patrol base. One is getting a haircut, one is giving a haircut, and the other two are just harassing them.

Boonies Haircut

Figure 58-3: I took this from the 'back seat' of a Huey, probably while I was flying with the Battalion CO as his RTO. The Huey was in a slight left turn when this picture was taken. When they turned really hard, if you looked out one door, you saw only the sky. If you looked out the other door, you were looking almost straight down at the ground. On your first flight, you worried about falling out the open door, but soon realized you wouldn't. After that, you didn't think about it at all, you just enjoyed the flight.

Huey—Passenger Eye View

Figure 58-4: Usually, Huey's flew in groups of three, in a triangle formation. This picture is of the Huey in the second row with the Huey I was in. The lead Huey is probably to the left and flying between us. Charlie Company, or elements of it, are being taken somewhere.

Flying Close

Figure 58-5: This shows a group of three choppers coming in to pick up troops for an air assault (also called a combat assault). The soldier on the ground is probably one of three soldiers (two not shown) signaling the landing spot for one of the choppers. As soon as they landed, troops would run over and board them from both sides. Groups of soldiers can be seen waiting on the far left and right of the picture. Once one group of three Hueys were loaded and off, another group of three choppers would come in for more troops.

Air Assault — Coming in for Pickup

Figure 58-6: This shows a group of three Huey's (the third is hidden in the dust or out of picture on the left) landing and discharging troops. I was already on the ground, having landed with a previous group of choppers. Normally the LZ for a combat assault would just be a naturally open area with short vegetation. I don't remember why this one was so dusty. No bad guys were waiting for us.

Air Assault — Landing in LZ

59. Loss After Gain

It was October 1st, 1967. I was working as an RTO in the battalion Tactical Operations Center (TOC) for the 1/12th, 4ID. The firebase was in an area west of Pleiku, near the Central Highlands.

That morning we received an excited call from a FAC (forward air controller) who was flying a few miles away from the firebase. He said that he could see a large force of NVA soldiers crossing an open grassy area. He said he was calling in air support but wanted to begin an artillery assault on the enemy below him until the aircraft arrived. He gave the coordinates and said he was moving out of the line of fire.

Our artillery immediately sprang into action, delivering volley after volley at the enemy. When the war birds arrived, the FAC requested a cease fire of the artillery so his aircraft could safely continue the attack on the enemy. When the artillery fire ended, the air attacks began. We watched the diving and strafing aircraft from the firebase and heard the explosions of their bombs. Those NVA were being pounded (See Figures 59-1 and 59-2).

This was a completely out of the norm situation. The NVA

rarely traveled by day and absolutely never through open grass-lands. To catch them making that huge mistake was a delight to everyone in a US uniform. They must have had a good reason to expose themselves like that, but we didn't care. When the aircraft were done, the artillery started up again. We kept it up for a couple of hours. Finally, they called a cease fire.

All the while the bombardment was going on, our battalion officers were planning what to do next. Our military intelli-gence officer, Captain Pat, decided to immediately go into the area via helicopter to see if he could gather any immediate in-telligence from the dead NVA's equipment.

Cpt Pat had been the head of battalion S-2 (Intelligence) for about five or six weeks. He was an extremely bright and a very likable guy. He was always smiling and joking with every-one — officers and enlisted guys alike — it didn't matter to him. He was from Pennsylvania. I had worked with him a lot and like everyone who did considered him a friend.

Soon, Cpt Pat, his RTO, and a couple of other guys jumped into a waiting helicopter and flew to the area. His RTO later told the story of what happened.

When the helicopter reached the scene of the bombing, they could see lots of NVA bodies scattered around a large area. Cpt Pat directed the helicopter pilot to land near a group of three or four bodies. He wanted to grab any packs or documents they might have and bring them back to the firebase to examine for any information they might contain. Just as the helicopter was about to set down, one of the 'dead' NVA rose up and, using his AK-47, shot several rounds at the landing chopper. One of the rounds struck Cpt Pat in the chest. The door gunner returned fire as the helicopter climbed back into the air. The NVA soldier fell.

The helicopter with the wounded Captain immediately flew to the nearest field hospital. Tragically, Cpt Pat died from his wound.

When the word came back to the firebase that we had lost Cpt Pat, the mood was very somber. As I said, everyone who had met or worked with him liked him very much. He was only in his mid-twenties when he died.

Everyone should have been ecstatic about the losses we inflicted on the NVA that day, but the death of Cpt Pat tempered everyone's feelings for several days.

Later that day, one of our infantry companies arrived at the area of the bombing and secured it. The equipment that the dead NVA carried was laying there with the bodies. They also started collecting everything the enemy had left behind or dropped as they tried to evade our artillery and aircraft. There was a lot of it. A couple of wounded NVA were captured as well.

The infantry company spent all that day and the next gathering NVA equipment. It was all brought back to the firebase. Several of us helped process the equipment that was brought in.

Dozens of AK-47s and SKS's, many grenades, and considerable ammo was collected. Two heavy machine guns mounted on wheels were also part of the captured weapons (See Figure 59-3). In addition, there were piles of personal equipment like shovels, hats, and canteens, and the packs all of it was carried in (See Figure 59-4). Lots of rice and the oil for cooking it and cooking pots were captured too. The rice was carried in large baskets and the oil in big bottles (See Figure 59-5). Everything was there that a sizeable NVA unit needed to make war on us. It was a very big haul.

The two wounded NVA soldiers were brought to the firebase for treatment as well (See Figure 59-6).

The battalion commander decided to show everything off to the brigade and division brass. To make it all look nice, many of us spent a couple of days cleaning the NVA weapons. Then, many of the weapons and other captured gear was laid out nicely for the upcoming dog and pony show. Four days after the bombings, our firebase was the site of that big dog and pony show for the division brass.

Two things were reinforced in my mind by this incident. First, the NVA were very good soldiers. They had no helicopters to bring in supplies to them daily like we did. Everything they needed to sustain themselves in a war zone had to be carried with them. They might get a little help from the civilians in the area, whether voluntarily or forced, but that was it. They definitely had courage and determination. They were formidable enemies and had my respect.

Second, I wanted revenge even more. Revenge for all the young lives we lost, for all the friends that were now gone, for Cpt Pat. I was never able to get that revenge during the rest of my time in Vietnam. It took me years to stop feeling that way.

Figure 59-1: A large group of NVA soldiers were spotted by a FAC, in the open, during the day, not too far from the firebase. The FAC called in an air attack on them. Here, a bunch of guys from the battalion command group are watching the attacking aircraft. The terrain around the firebase was mostly sparsely treed grasslands so the NVA were easily seen.

Watching Air Attack

Figure 59-2: Two jets making their attack run, flying right over the firebase. The air attack went on for, probably, an hour and was followed up by artillery attacks. We really poured it on those NVA soldiers caught in the open.

Two Jets Attacking

Figure 59-3: A few of the captured NVA weapons. On the left are AK47s, on the right are SKSs and an RPG rocket head. In the foreground are two heavy duty machine guns mounted on wheels. These machine guns had to be pulled through the grass and wooded areas. That couldn't have been fun duty. Odds are the intended use for these weapons was for an attack on our firebase. We were glad they didn't have a chance to use them on us.

Captured NVA Weapons

Figure 59-4: Some of the equipment that was thrown away as they tried to escape or belonged to dead NVA. A little bit of everything.

Captured NVA Equipment

Figure 59-5: Part of what was found after the attack on an NVA unit caught in the open was over. This shows some of the food they were carrying. All the captured material was examined by our intelligence group.

Captured NVA Food

Figure 59-6: One of the wounded NVA soldiers who was unable to escape with his comrades. He is being treated by one of our medics.

Captured NVA Soldier

60. Almost Forgotten

I was getting to be a real short timer. My rotation home date was 10/13/67, the day my one-year tour in Vietnam was over. I only had 10 days to go.

In our unit at that time the regular practice was when a soldier in the field had about 10 days left in his tour, he was transferred back to the 4ID base camp. The main purpose for that was to lessen the chance that a soldier would be killed in his last 10 days. The 4ID base camp was a significantly safer place to be compared to the boonies, including the firebase.

That morning, I waited patiently for the order to pack my gear and get on that chopper bound for base camp. But no one said anything to me.

The next day was the same. No order to go back.

Then I had eight days to go. No one told me to pack up.

I began to wonder why.

I felt I had a pretty important job in the TOC. I was the guy scheduling all the radio watches and taking care of the daily operational duties. I started to think that they were just trying to decide on a replacement for me and didn't want to send me back till they found one. Still, I wondered, why they didn't say anything?

This continued on the same way day after day, until I only had three days left in Vietnam. I was finally starting to get a bit irritated that no one said anything to me about my delay in going back to base camp.

Late that morning, I approached the battalion Sgt Major, who I knew pretty well. I tried to be casual when I asked, "Top, do you know when I will be going back to base camp?" He looked at me with a puzzled look on his face and asked, "Why?"

"Well", I said, "I'm due to go home in three days." His eyes opened wide. "Is that right?" he asked. "Yes, Top." His immediate response was, "Get your shit packed. You're going back this afternoon."

I went back to my bunker, gathered my few personal belongings, and put them in my rucksack (See Figure 60-1). Then, I took all but two of my ammo magazines and all my grenades over to a bunch of guys on the line and said they could take what they wanted. I kept the two magazines, just in case.

In early afternoon, Top came to my bunker and told me to come with him. We walked together over to the TOC. There a gathering of the officers and enlisted men I had worked with for several months stood waiting for me. The battalion commander came over to me. First, he thanked me for my service and the job I had done while working in the TOC. He wished me luck in the future and a good trip home. Then he did something that I have always considered to be quite an honor. He gave me a sandbag.

It had been an established tradition in our battalion that whenever an officer was transferred out of our battalion, whether going home or just to a different unit, he was given a Red

Warrior souvenir sandbag. The sandbag was painted with a sil-houette of an Indian chief's head with "Red Warriors" printed above it and "Vietnam" printed underneath (See Figure 60-2). They gave one of those sandbags to me, a 'lowly' SP4. I'm pretty sure my eyes teared up. I thanked all of them, but I don't think I could get much else out.

After that small ceremony and a bunch of handshakes, I went back to my bunker to get my stuff and took my last picture in the boonies with Susie (See Figure 60-3).

I then went to the LZ to wait for the chopper that would start me on my journey home. Top walked me to the LZ. On the way, he apologized for the paperwork mix up. He was sorry I was almost forgotten. He said that it was lucky I said some-thing because there was no telling how long it would have taken for someone else to notice it. I thought to myself there was no chance that I would have forgotten about my return home date.

When the chopper arrived, I walked over, got on, and it lift-ed off. As we flew up and over the firebase, I looked down on it and felt.... guilt. I felt guilty leaving all my brothers behind. I knew I shouldn't feel that way, but I couldn't help it. All I could do was wonder what the future held for all of them. As I flew away, in my head, I simply thought, "Good luck, guys."

Back at base camp, when I got to the company area I was told that the Sgt Major had radioed in and given orders that I was to get expedited processing. I flew out of the 4ID base camp the next day and actually left Vietnam a day earlier than my DEROS date. Thanks, Top.

Figure 60-1: This was taken as I packed my gear just before starting my trip home from Vietnam. I gave most of the ammunition and grenades I always kept with me, to anyone who wanted to add a little more to his own personal arsenal.

Packing Up to Go Home

Figure 60-2: This is the souvenir sandbag given to me by the battalion commander on my last day in the field before my return home. This bag was normally only given to officers when they left the Red Warriors battalion. I felt very honored when they presented it to me as appreciation for my good work in the TOC.

Souvenir Sandbag

Figure 60-3: Just before I walked over to the LZ to board the helicopter for my trip back to the 4ID base camp, I called Susie over and gave my camera to a friend. I told him to take the last picture of me in the boonies of Vietnam with my special friend, Susie.

Last Picture in the Boonies

61. The Trip Home

Vietnam, Mid-October 1967, 4ID base camp.

I had just returned to the 4ID base camp outside of Pleiku, Vietnam from the firebase for the 1/12th. I was due to rotate home from Vietnam in three days. Normally guys came back to base camp about 10 days prior to DEROS date. Because of some mix up with paperwork, I was about seven days late coming back from the boonies.

The afternoon I got back to base camp I reported in to the company office. When I walked in and told them who I was they told me they were expecting me. The battalion Sergeant Major had radioed that I was coming and that my paperwork was to be expedited. The guys in the that office must have thought I was somebody important for that to happen.

The Sgt Major was just trying to make up for the foul up in the paperwork which kept me out in the boonies much longer that I should have been out there. Our unit didn't want anyone killed or wounded in his last 10 days in-country, so that was the reason for the 10-day rule. Base camp was much safer than anywhere out in the boonies, even the firebase.

I did get that expedited processing. In fact, the next day I was on a cargo plane that took me to, I believe, Cam Ranh Bay airfield. I stayed there overnight and the next morning all of us guys returning to the States were bused to the airport.

Once there, we all had to go through a long line to get our baggage inspected. I guess the military didn't want us taking home any unauthorized items. You know, like automatic weapons, grenades, C-4 explosive, etc. The only thing we were supposed to have in our bags was clothing and some toiletries.

The baggage inspection room had a long line of tables set up in the middle of the building. On one side of the table were a few NCO's who inspected our bags. We walked down the other side of the tables, pushing our open bag along on the table. As we walked along the tables, the soldiers on the other side rummaged through our stuff. I knew I had nothing in my suitcase that was not allowed. Or so I thought.

When I got to this one buck sergeant, he saw a sandbag folded up in my suitcase. He grabbed it and threw it behind him on the piles of other stuff that was not allowed to be taken home. Very gruffly he said, "Those are not allowed to leave Vietnam."

That sandbag was a commemorative sandbag given to me by the battalion commander when I left the firebase. It was not just a common sandbag. I tried to explain to him that it was a special sandbag and he just told me to keep moving. I did, but when I reached the end of the inspection line I just waited.

Luckily, I was near the end of the line of soldiers getting bags inspected, so not too many guys were behind me. I waited until the last guy went through. Just then, the a-hole Sgt that took my sandbag apparently decided he needed a break and left the building. I went back to the spot where my sandbag was laying on the floor against the wall. Unfortunately, there was a line of tables between me and it.

Just then, another Sergeant came into the area behind the tables. I asked him to come over. I pointed to my sandbag on the floor and explained why I had it and what it was. I asked him to just look at it. He went over, picked it up, unfolded it, and saw the painting of the Indian head, the 'Red Warriors' written above it and 'Vietnam' written below it. It was obviously not just an ordinary sandbag. The sergeant just shook his head and said, "Here you go." and gave it back to me. I thanked him very sincerely and went outside and got on the bus waiting to take us to the tarmac.

On the tarmac at the airport, a civilian airliner, Seaboard World Airlines, was waiting for us. We got off the bus and climbed up the steps to board the plane. Waiting inside were several attractive stewardesses. Everyone just took the first seat available (See Figure 61-1).

The conversations in the plane were subdued as it taxied toward the runway. It was even quieter when it started revving up its engines to takeoff, and quieter still as it started to race down the runway. But, just as it lifted into the air, there was a loud, spontaneous cheer from every GI on board. Fists were raised in the air and guys were shaking hands and clapping. That takeoff, it seems, was the symbolic sign to every single GI onboard that he had made it through Vietnam and was actually going home. And thanks to the Sgt Major, I was going home one day early!

One year before, as I said goodbye to my family, I had the deep feeling I was not going to come home from Vietnam. As that plane took off, I realized I had been wrong. I was going home. Then I realized I now had to figure out what I was going to do for the rest of my life. It was an odd feeling.

The flight back to the States was uneventful. Unlike the C-141 I flew in to get to Vietnam, this aircraft had lots of windows and food service. And stewardesses! Nice.

We flew non-stop into Fort Lewis, Washington where we were processed back into the States. They also served us a nice steak dinner there. After processing and food, there were buses waiting to take us to the Seattle Airport.

Before leaving Fort Lewis, most of us changed from our Khakis into civilian clothes. I changed into the civies that I had brought to Vietnam one year before and had stored for a year in a locker at base camp. Everyone was so anxious to get to the airport that a lot of guys didn't bother packing their Khaki uniforms back into their suitcase. When we left, Khakis were laying all over the floor of the big room in which we changed.

I was heading home to Kenosha, WI, which is about midway between Milwaukee and Chicago. At the Seattle airport, I checked on stand-by flights to Milwaukee or Chicago. I didn't care which I went to. The first flight available was to Chicago, so I signed up for it. It was a non-stop and left about 7:00 PM. I made it on-board with my stand-by ticket.

On the flight to Chicago, I just happened to sit next to a gentleman who worked for IBM. One of my best friends had just gone to school to become a computer programmer. In his letters to me in Vietnam, he encouraged me to investigate that career too when I got home. I told my seatmate about that, and he also encouraged me to check into it. He tried to explain what a computer programmer did but most of it went way over my head. Still, he was very upbeat about the prospects of having a career in computer programming. I did follow the advice I received from both. After I got out of the Army, I went to programming school and ended up having a great, 45-year career as a computer programmer.

When we landed in Chicago, it was about 1:00 AM. I still needed to get to Kenosha. I thought I'd check to see if there was a train running between Chicago and Kenosha at that time of night. I looked for a phone booth. I found one, sat down inside it, and almost freaked out. I thought I'd entered a time warp or something, because the pay phone in front of me didn't have a rotary dial, it had buttons! Holy crap, how long had I been gone? I had never even heard of a phone that only had buttons.

I looked up the phone number for the train station in the phone book. Then, I took a deep breath, put in a dime, and started pushing buttons to make the call. As I did, each button I pushed played a different musical beep in the earpiece. Wow. Once I connected to the train station, I was relieved that the phone worked the same as the phones I had left behind so long ago. I learned there was a train that left Chicago for Kenosha in a couple of hours. I took a cab to the train station and bought a ticket.

The train arrived in Kenosha about 4:30 AM. I got off and was surprised and very happy to find there was one cab waiting outside the station. My family ran a bakery in Kenosha, and I knew my dad was working at that time of night, so I figured I'd go there.

One note I want to add here. On my trip home, I encountered no, nor did I see any disrespect for any soldier on the journey.

When I got into the back seat of the cab, the cabbie had a buddy in the front seat keeping him company. On the ride to the bakery, his buddy was telling the driver how tired he was of living in Kenosha and how he was on the verge of leaving this town forever. In his words, "This town sucks!" Here I was, at that moment, the most grateful person in Kenosha to be back there, and here was this other guy saying how much he wanted to leave. Life is just a collection of different points of view, isn't it?

In my letters home over the previous month, I had never given my family any specific date they should expect me home. I didn't want to worry them if I didn't make it on that date. I just told them I thought I'd be home sometime in the middle of October. So, it was a complete surprise to everyone when I walked in the back door of the bakery and yelled, "Hi Everybody!" Beside my Dad, there were seven or eight people working there that night, and I knew every one of them. They were all like family to me. It was a great way to start my homecoming, with back slaps, hugs, and handshakes from all of them.

After talking at the bakery for an hour, it was about 6:00 AM and time to head home to see my Mom, my younger sister, and two younger brothers. My Dad and I left the bakery and drove the 10 minutes to my parents' home. We went in and as I climbed the stairs to where the bedrooms were, I loudly yelled, "Rise and shine, everyone. Dennis is home!" Four pairs of feet all hit the floor at the same time, and the rest of my family came running out of their bedrooms.

Lots of hugs and kisses later, I was officially home from Vietnam.

Figure 61-1: Taken just before boarding the plane that was to take me back to the States. I was probably saying to myself, "I don't believe I'm actually going home." I imagine I was not alone thinking that in the group of soldiers climbing the staircase.

Homeward Bound

Glossary

1/12th: 1st Battalion, 12th Infantry Regiment - The Red Warriors battalion

105: 105mm Howitzer—Artillery cannons which were at the 1/12th fire base

4ID: 4th Infantry Division

4.2: 4.2-inch mortars—Usually were only at the firebase

50-cal: .50 caliber machine gun—Heavy machine gun, usually mounted on a vehicle

81: 81mm mortar—The mortars usually carried by the troops in an infantry company

Air Assault: Moving troops into an area using helicopters

AIT: Advanced Infantry Training—Two-month school following basic training. Specifically for training infantrymen.

AK-47: Prime weapon of the NVA and Viet Cong—Fully automatic

Alpha: The letter 'A' phonetically—Nickname for 'A' Company

AO: Area of Operations—A section of country, usually about 6 to 7 miles wide and long in which a battalion and its companies operated search and destroy missions

APC: Armored Personal Carrier—Tracked vehicle designed to carry 10 to 12 troops

ARVN: Army of the Republic of Viet Nam

B-52: Heavy Air Force bomber

Base Camp: The primary headquarters area for a division—very secure area

Battalion: Three to five infantry companies, each composed of 100+ soldiers, plus support troops. Each company is named with an alphabetic letter.

Boobie Trap: Hidden weapon designed to hurt or kill the enemy by surprise

Boonies: Anywhere in Vietnam outside of base camp

Bravo: The letter 'B' phonetically—Nickname for 'B' company

Brigade: Military unit that is composed of one or more regiments

C-130: Medium Air Force cargo plane — propeller driven

C-141: Heavy Air Force cargo plane - Jet

C-4: Plastic explosive — Explodes with a blasting cap, burns with a match

C-47 (AC-47): Air Force plane converted to a gun ship — Nicknamed Puff or Spooky

Call Sign: The code name for whoever is talking on the radio

Captain: Rank of the officer that usually commands a company

Care Package: Any package we received from home that contained goodies (food) or other things we needed

Charlie: The letter 'C' phonetically — Nickname for 'C' company. It was also a nickname for the VC or NVA

Chinook: Heavy, two rotor, cargo helicopter

Claymore mine: An antipersonnel mine carried by the infantry which, when detonated, propelled small steel cubes in a 60-degree fan-shaped pattern to a maximum distance of 100 meters

CO: Commanding Officer of a unit — Company, Battalion, Brigade, etc.

Command Bunker: Bunker for the commanding officer, senior NCOs, and support personnel

Commander : Leader of a military unit

Company: Four platoons of soldiers, each composed of 20 to 50 soldiers

Contact: Term for describing when friendly forces engage the enemy face to face—usually resulting in a fire fight with small arms

CP: Command Post—the area where the unit command group reside

C-Rations: Canned meals. Generally, 12 different meal combinations were available in a single case. Each meal provided approximately 1,200 calories and was composed of multiple entrees

Def-con: Defensive concentrations—pre-targeted artillery points around a position held by friendly forces

DEROS: Date of Estimated Return from Overseas Service—the date you were going home.

Deuce-and-a-half: Army truck with a two-and-a-half-ton load limit

Dust-off: Nickname for the medical helicopters that took wounded personnel to treatment facilities. They were partially manned by soldiers with medical training to render first aid to the wounded as they were transported

Echo: The letter 'E' phonetically—Nickname in the call sign that identified who the RTO worked for

F-4 Phantom: Fighter / Bomber jet commonly used in Vietnam

FAC: Forward Air Controller—liaison between the fighter/ bomber aircraft and the ground troops. Flew at low altitudes, in slow aircraft, to better see the battle zone

Fatigues : The clothing worn by troops in the field. Jungle fatigues had more pockets and a looser fit than standard fatigues

Firebase: The position in the field which housed the battalion command and artillery

Firefight: A battle, or exchange of small arms fire with the enemy

Flares: Illumination projectile. Hand-fired, shot from artillery or mortars, or dropped from aircraft. Generally supported by a parachute

Flank: Either side of a military formation

FO: Forward Observer—A member of the Artillery group, generally inserted with the infantry to coordinate supporting artillery fire

Free fire zone: Combat areas in which there are no civilians, only enemy soldiers. The rules of engagement are, if it not a GI, shoot.

Friendly Fire: Any type of weapon accidentally aimed at friendly troops and fired by friendly troops.

Grunt: Infantryman

H & I: Harassment and Interdiction—Artillery or any heavy weapons fired at random times at suspected enemy locations or areas of movement

HE: High explosive—generally as artillery and mortar rounds

HHC: Headquarters Company—the command and support personnel of a battalion

Hooch: Overnight shelter built by connecting two ponchos and supported by sticks generally obtained from the surrounding area

Hot LZ: Landing Zone that has enemy soldiers nearby who are shooting at landing helicopters and personnel

Huey: Nickname for the UH-1 helicopter used extensively in the Vietnam War

Hump: Moving through rough terrain, usually loaded down with heavy gear

Incoming: Any kind of weapon(s) shot toward you or your position

Jungle Boots: Lightweight boots issued to troops in Vietnam with cleated soles and nylon and leather uppers

KIA: Killed in Action

Kilometer: 1,000 meters — the normal military measurement standard

Klick: A kilometer

Landing Zone: A natural or manmade area, clear of obstacles, that helicopters can land in

LP: Listening Post — usually a group of 3 to 4 soldiers who station themselves outside a defensive position during the night to provide early warning against a surprise attack

Lit-up: Heavily shot by any kind of weapon

LT: Lieutenant — Lowest rank of officer — generally platoon leaders

LT Colonel : Rank of the officer that usually commands a battalion

LZ: Landing Zone

M-14: .30 caliber semi-automatic rifle generally used by the military before the Vietnam War

M-16: .223 caliber fully automatic rifle use by the military through and after the Vietnam war

M-60: Squad size .30 caliber belt fed machine gun carried by soldiers in Vietnam and later conflicts. Usually about eight were in each company

M-79: Single shot, grenade launcher that shot 40mm grenades

Mad minute: Short period of time when all the soldiers of a unit could fire their weapons simultaneously to test the function of the weapons

Mama san: Nickname for an older Vietnamese woman

Meter: Standard unit of measurement in the military—about 39 inches

MIA: Missing in Action

Mini gun: A multi-barreled machine gun capable of firing up to 6,000 rounds per minute. In Vietnam, it was typically mounted on a C-47 gunship commonly called Puff or Spooky

Montagnard : Member of a tribe of mountain people in Vietnam

Mortar: A muzzle-loading cannon with a short tube in relation to its caliber that throws projectiles with low muzzle velocity at high angles.

MOS: Military Occupational Specialty—The job a soldier was trained to do

Napalm: Highly inflammable jellied petroleum bombs usually dropped by aircraft

NCO: Non-Commissioned Officer—enlisted men of higher rank

NVA: North Vietnamese Army—Regular, trained soldiers that invaded South Vietnam

OP: Observation Post—usually a group of 3 to 4 soldiers who station themselves outside a defensive position during the day to provide early warning against a surprise attack

P-38: Tiny, handheld, collapsible, metal can opener used to open C-Ration cans

Papa san: Nickname for an older Vietnamese man

Palace Guard: Nickname for providing security at a firebase by an infantry company

Patrol Base: A nighttime perimeter where we stayed for more than one night so we could search an area more thoroughly and rest

Piss-tube: An empty cardboard artillery shell tube, open on both ends, and stuck in the ground. Used to provide sanitary urination facility for soldiers at a firebase

Platoon: A group of between 20 and 50 soldiers. Usually composed of three or four squads which are each made up of two or three teams

Point: The lead element of a formation of soldiers on the move. Usually positioned about 75 meters in front of the main body.

Poncho: A square shaped, waterproof nylon rain gear. It has a hood in the middle and the rest just drapes over the body

Poncho liner: Lightweight nylon, camouflaged, blanket

POW: Prisoner of War

PRC-25: Radio carried by an RTO while a military unit is on the move

PTSD: Post Traumatic Stress Disorder—Mental illness caused by exposure to very stressful conditions such as combat

Puff: Nickname for a C-47 gunship in Vietnam—also called Spooky

Quad 50: Four .50 caliber machine guns mounted on the same platform designed to fire simultaneously. Usually truck mounted.

R & R: Rest and Relaxation — nickname for the five day 'vacation' provided by the military for military personnel in Vietnam. There were multiple destinations a soldier could go to, among them Hong Kong, Bangkok, and Taipei

Red Warriors: Nickname for the 1st Battalion of the 12th Infantry Regiment

Regiment: A Regiment is composed of multiple battalions

RPG: Rocket Propelled Grenade — Shoulder fired grenade meant for attacking tanks, helicopters, or bunkers

Rotate: Nickname for being transferred from Vietnam back to the States

RTO: Radio Telephone Operator — Soldiers who carried and / or operated a radio

Rucksack: Backpack issued to infantry in Vietnam — a metal frame with a nylon pack attached

S-2: Intelligence section — Gathered and interpreted intelligence

S-3: Operations section — Handles planing, operations, and training

S & D: Search and Destroy — Operation whereby an infantry unit moved through the landscape searching for the enemy and, once found, strove to destroy that enemy

Saddle Up: Term meaning get your gear on and prepare to move out

Short / Short timer: A soldier that has only a limited time left in his tour before being sent home. The time period remaining might be only a few days or a few weeks depending on how the soldier interpreted it in his mind

Shrapnel: Pieces of metal—large and small - sent flying by an explosion

Slit Trench: A narrow ditch in the ground that can be used for latrine facilities.

Sit-rep: Situation Report—The act of reporting to another entity your status currently—normally by radio. Usually, the report is only to ensure to others that your status is normal.

Six: The radio call sign of the leader of a unit. E.g., the company commander

SKS: Secondary weapon of the NVA and Viet Cong—Semi-automatic rifle

Starlight Scope: Early light enhancement device—similar in size to a large hand-held telescope

Steel Pot: Nickname for the helmet worn by soldiers. Composed of two parts. An inner, plastic liner that contains the webbing that fits it on the head and a steel protective outer layer that fits snugly on the inner plastic layer

Squad: A group of eight to twelve soldiers

Squelch: The sound made by an incoming radio transmission when the sending radio's talk button is released

Team: A group of three to six soldiers

Tet: Lunar New Year—In Vietnam it was often abused by the NVA and Viet Cong when a 'cease fire' was negotiated and the NVA and Viet Cong didn't honor it

TOC: Tactical Operations Center—two trailers that contained radio and other communication equipment that served as the communication center for a battalion at a firebase

Top: First Sergeant—most senior NCO in a Company or Sergeant Major—most senior NCO in a battalion or higher—a universal nickname

Tracer: Bullet chemically treated to glow after it is fired from a weapon so its flight path can be easily seen

Trip flare: A small ground flair designed with an attached wire that can be set up as a hidden booby trap to alert defenders to enemy soldiers moving through the darkness

Twin 40: Two 40mm cannons mounted on a tracked vehicle similar to a tank. The two 40mm cannons are typically fired in an alternating manner at a rate of 260 rounds per minute

USO: United Services Organization—provided entertainment to the troops and designed to raise morale

VC / Viet Cong: Nickname for the Vietnamese communist sympathizers who fought against both the Vietnamese government and all its allies

Weapons Platoon: Infantry soldiers trained to use the mortars we carried with us

Web gear: Canvas belt and suspenders for carrying equipment and ammunition on infantry operations.

WIA: Wounded in Action

Willie Peter / WP : White Phosphorous—Grenade or shell containing white phosphorous, a chemical which burned very hot after the device exploded. When white phosphorus hit the skin of a living creature it continued to burn until it had burned through the body. Water would not extinguish it

About the Author

Dennis Witt served as an infantryman in Vietnam in the 4th Infantry Division from October 1966 till October 1967. During his tour he was awarded two Bronze Stars, one for valor and the other for meritorious service.

Since 2003, he has been active in the Red Warriors Vietnam Association which is composed of soldiers that served in his battalion during the five years it was in Vietnam.

He was born in Minnesota but grew up and lived most of his life in Wisconsin.

After the service he worked for 45 years as a mainframe computer programmer and systems analyst. He worked for the same company for the duration of his career.

He has been married to his wife, Kathy, for over 50 years and has two children and two grandchildren. Dennis enjoys his family, traveling, hunting, fishing, camping, and outdoor sports.

He wrote *Infantry Life: Vietnam's Central Highlands 1966–1967* to describe for family and friends what daily life was like for infantrymen serving with the Red Warriors battalion in Vietnam in 1966 and 1967.